Labo...

Physical Geol...

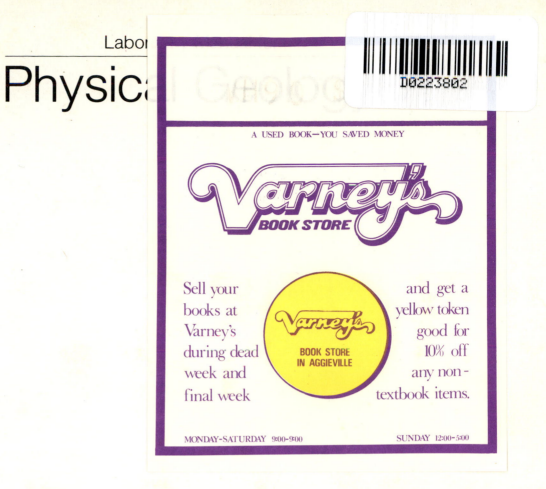

Laboratory Manual for

Physical Geology

Seventh Edition

James H. Zumberge
Professor of Geosciences and President,
University of Southern California

Robert H. Rutford
Professor of Geology and President,
University of Texas at Dallas

wcb

Wm. C. Brown Publishers
Dubuque, Iowa

Book Team

Editor *Jeffrey L. Hahn*
Developmental Editor *Lynne M. Meyers*
Designer *David C. Lansdon*
Production Editor *Gloria G. Schiesl*
Photo Research Editor *Carol M. Smith*
Visuals Processor *Joyce E. Watters*

wcb group

Chairman of the Board *Wm. C. Brown*
President and Chief Executive Officer *Mark C. Falb*

wcb

Wm. C. Brown Publishers, College Division

President *G. Franklin Lewis*
Vice President, Editor-in-Chief *George Wm. Bergquist*
Vice President, Director of Production *Beverly Kolz*
National Sales Manager *Bob McLaughlin*
Director of Marketing *Thomas E. Doran*
Marketing Information Systems Manager *Craig S. Marty*
Marketing Manager *Matt Shaughnessy*
Executive Editor *Edward G. Jaffe*
Manager of Visuals and Design *Faye M. Schilling*
Manager of Design *Marilyn A. Phelps*
Production Editorial Manager *Julie A. Kennedy*

Cover Image: © N. Foster/The Image Bank. Erosion by running water has carved these colorful sedimentary strata in the Canyonlands National Park in Utah. Sediments derived from this erosive action eventually reach the Colorado River flowing through the region and are transported to some distant site of deposition. The desert climate supports only sparse vegetation, which accounts for the virtually continuous rock outcrops seen in this photograph.

Illustrations: Norm Frisch, base art for figures 1.52, 1.53, 1.54, 1.55, 1.56, 1.57, 2.4, 3.22, 3.23, 3.44, 3.49, 3.54A, 3.54B, 4.2, 4.11, 5.3. Gary Port, color application for figures 1.49, 1.52, 1.53, 1.54, 1.55, 1.56, 1.57, 3.18, 3.30, 3.38, 4.1, 4.3, 4.4, 4.5, 4.6, 4.7, 4.8, 4.9, 4.10, 4.11, 4.22, 4.23, 4.24, 4.25, 4.26, 5.1, 5.3. Larry Bowring, cartography for figure 3.43.

Library of Congress Catalog Card Number: 87-71346

ISBN 0-697-05097-1

Printed in the United States of America by Wm. C. Brown Publishers
2460 Kerper Boulevard, Dubuque, IA 52001

10 9 8 7 6 5 4 3 2

Contents

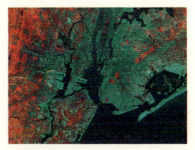

Aerial Photographs and Other Imagery from Remote Sensing 60

Part 3
Geologic Interpretation of Topographic Maps, Aerial Photographs, and Earth Satellite Images 69

Part 4
Structural Geology 143

Part 5
Plate Tectonics and Related Geologic Phenomena 179

Materials Needed by Students Using This Manual

1. Scale ("ruler") graduated in tenths of an inch.
2. Colored pencils (red, blue, and assorted other colors).
3. Felt tip pens ($\frac{1}{8}''\times\frac{1}{4}''$ tip), three assorted colors.
4. Several medium to medium-soft pencils (2H or No. 2).
5. Pocket stereoscope.
6. Small magnifying glass for map reading.
7. Six sheets $8\frac{1}{2}''\times 11''$ tracing paper.
8. Eraser (art gum or equivalent).
9. Inexpensive pencil sharpener.

Preface

Since the first edition of this Manual was published in 1951, the science of geology has undergone a revolution unprecedented in this century. Space imagery has expanded our ability to see our planet in a new and broader perspective. The subject of plate tectonics has brought together a number of heretofore unrelated geologic events and has provided a grand scale for the study of fundamental processes that are as operative today as they were in the geologic past.

Yet, we need to remember that some of the basic elements of earth science remain as valid today for elementary students as they were in mid-century. The study of minerals and rocks at the elementary level remains more or less unchanged, and topographic maps and aerial photographs are still fundamental tools in the study of landforms produced by the various geologic processes. Structural geology remains unique to the geological sciences, and must be included in a workbook at the elementary level.

In preparing this seventh edition of *Laboratory Manual for Physical Geology,* we have received considerable help from those who have used the sixth and earlier editions. Their experience in dealing with students in the laboratory has been passed to us through comments and expanded reviews requested by the publisher. Many of their ideas have been incorporated in this edition.

Among these is a new section in Part 1 that follows rock identification. This segment on the occurrence of rocks introduces students to the concept of geologic time and the relative ages of rock masses. We lead the students through a presentation that explains how a geologic column can be derived from the relative ages of formations as depicted on uncomplicated and simplified geologic cross-sections. We believe that this early presentation of the occurrence of rocks in the field, as opposed to hand specimens, will get students thinking in the third dimension and about the magnitude of geologic time, two thought processes that are the most challenging for beginning students.

Part 2 introduces the student to topographic maps and aerial photographs and some space imagery, and stresses their importance in identifying the landforms produced by the various geologic agents.

Part 3 deals with the geomorphic interpretation of topographic maps and aerial photographs. Some of the exercises in this part remain more or less unchanged from previous editions, but many have been revised extensively; and some are new. The organization of Part III has always been a problem because the order in which the geologic processes are presented in courses in physical geology differs from institution to institution; nevertheless, we believe that instructors using this Manual will select exercises in an order commensurate with their own system of presentation without too much difficulty.

Part 4 is devoted to structural geology with special reference to the geometry of folds and faults with emphasis on the relative age relationships of formations shown on geologic maps, aerial photographs, or block diagrams.

Part 5 is essentially new and introduces the student to some of the aspects of plate tectonics, including the relationship of earthquakes to boundaries of spreading plates and subduction zones. Students are asked in exercises to calculate spreading rates along a segment of the East Pacific Rise, and movement of volcanic islands across the Hawaiian hot spot. The origin of atolls is introduced with reference to Darwin's Theory of sinking volcanic islands, and illustrated with new photographs of atolls in French Polynesia.

The addition of full color to this edition enhances the presentation of diagrams and maps. We still rely heavily on black and white aerial photographs, however, since they are still a standard tool in geologic mapping and geomorphic interpretation.

We are especially grateful to many of our colleagues who have supplied data and guided us to materials needed in the construction of new exercises and revisions of older ones. We acknowledge, with thanks, the help from George Gryc, Mark Meier, and Andrew Fountain. Jeffrey Scovil did most of the photography on the mineral and rock hand specimens, and a few were done by Wade McCoy. Erwin Christian supplied the photographs of atolls. We are especially indebted to Carla W. Montgomery for the use of some of the illustrations in *Physical Geology,* 1987, Wm. C. Brown Publishers.

Also, to those who reviewed the sixth edition, we express our thanks and appreciation for their critical comments and suggestions for improvement, based on their own pedagogical experiences and philosophies. These include Lauren Brooker, Eric J. Klingel, Mervin Kontrovitz, Stephen W. Lenhart, William W. Locke, and Merlin J. Tipton.

As authors, however, we accept full responsibility for errors that have crept inadvertently into these pages, and we welcome comments from users, if such errors are discovered.

We express thanks to Mabel Simmons for her expert help in preparation of the manuscript. And finally, to the many professional men and women at Wm. C. Brown Publishers, we extend our gratitude for their design of the format and expert editorial help in transforming our manuscript and art work to this finished product.

James H. Zumberge
Los Angeles, California

Robert H. Rutford
Dallas, Texas

Earth Materials

Introduction

Materials that make up the crust of the earth fall into two broad categories, minerals and rocks. Minerals are elements or chemical compounds that are formed by a number of natural processes. Rocks are aggregates of minerals or organic substances that occur in many different architectural patterns over the face of the earth.

The first goal of Part 1 is to introduce beginning students of geology to the identification of minerals and rocks through the use of simplified classification schemes and identification methods. Students will be provided with samples of minerals and rocks in the laboratory. These samples are called *hand specimens*. Ordinarily, their study does not require a microscope or any means of magnification because the naked or corrected eye is sufficient to perceive their diagnostic characteristics. A feature of a rock or mineral that can be distinguished without the aid of magnification is said to be *macroscopic* (also *megascopic*) in size. Conversely, a feature that can be identified only with the aid of magnifiers is said to be *microscopic* in size. The exercises that deal with the identification of minerals and rocks in Part 1 are based only on macroscopic features.

The second goal of this part is to acquaint the student with modes of occurrence of rocks in nature and their relative age relationships. Many students using this manual will participate in organized field trips to observe firsthand how rocks occur in nature. To prepare students for this field experience, some basic geological principles on the mode of occurrence of rocks will be introduced in the form of simple geologic diagrams. The concept of geologic time will be explored and the relative age relationships of rock masses will be examined according to some basic geologic principles and assumptions.

Even if organized field trips are not a required part of your geology course, an understanding of these geologic diagrams will enhance your appreciation of rock strata and other geologic phenomena as you encounter them in your travels.

Minerals

Definition

A mineral is a naturally occurring, crystalline, inorganic, homogeneous solid with a chemical composition that is either fixed or varies within certain fixed limits, and a characteristic internal structure manifested in its exterior form and physical properties.

Mineral Identification

Because minerals are elements or chemical compounds, one way to identify them would be to run chemical analyses on all specimens to be identified. This is neither practical nor necessary, because other simpler and quicker means are available. These means employ the physical properties of minerals as the basis for identification. For example, the common substance, table salt, is actually a mineral composed of sodium chloride, and bears the name *halite*. The taste of halite is sufficient for recognizing it and distinguishing it from other substances that have a similar appearance, such as sugar (not a mineral).

Common minerals are recognized or identified by testing them for general or specific physical properties. The taste test applied to halite is restrictive because halite is the only common mineral that can be so identified. Other common minerals can be tested by visual inspection for the physical property of color, or by using simple tools such as a knife blade or a glass plate to test for the physical property of hardness.

The first step in learning how to identify common minerals is to become acquainted with the various physical properties that individually or collectively characterize a mineral species. These are described in the following paragraphs.

Properties of Minerals

The *physical properties* of minerals are those that can be observed generally in all minerals. They include such common features as color, luster, hardness, and specific gravity. *Other properties* are those found in only a few minerals. These include taste, odor, magnetism, and chemical reaction with acid. In your work in the laboratory use the hand specimens sparingly when applying tests for the various properties.

General Physical Properties

LUSTER. The appearance of a fresh mineral surface in reflected light is its luster. A mineral that looks like a metal is said to have a *metallic luster*. Minerals that are *non-metallic* are described by one of the following adjectives: *vitreous* (having the luster of glass); *resinous* (having the luster of resin); *pearly; silky; dull* or *earthy* (not bright or shiny).

Your laboratory instructor will display examples of minerals that possess these various lusters.

COLOR. The color of a mineral is determined by examining a fresh surface in reflected light. Color and luster are not the same. Some minerals are clear and transparent, thus colorless. The color or lack of color may be diagnostic in some minerals, but in others, the color varies due either to a slight difference in chemical composition or to small amounts of impurities within the mineral (fig.1.1).

STREAK. The color of a mineral's powder is its streak. The streak is determined by rubbing the hand specimen on a piece of unglazed porcelain *(streak plate)*. Some minerals have a streak that is the same as the color of the hand specimen; others have a streak that differs in color from the hand specimen. The streak of minerals with a metallic luster is especially diagnostic.

DIAPHANEITY. The ability of a thin slice of a mineral to transmit light is its diaphaneity. If a mineral transmits light freely so that an object viewed through it is clearly outlined, the mineral is said to be *transparent*. If the object viewed is not clearly outlined, the mineral is *translucent*. Some minerals are transparent in thin slices and translucent in thicker sections. If a mineral allows no light to pass through it, even in the thinnest slices, it is said to be *opaque*.

Figure 1.1 The specimens in this photograph are all quartz. The difference in colors is due to various impurities. Clockwise from the left: smoky quartz, amethyst, smoky quartz, rose quartz, quartz crystal. (Photo by Bob Coyle/WCB.)

Table 1.1. Mineral Hardness according to the Mohs Scale *(A)* and Some Common Materials *(B)*

HARDNESS	A	B
1.	TALC	
2.	GYPSUM	
2.5		FINGERNAIL
3.	CALCITE	
3.5		COPPER PENNY
4.	FLUORITE	
5.	APATITE	
5–5.5		KNIFE BLADE
5.5		GLASS PLATE
6.	ORTHOCLASE	
6.5		STEEL FILE
7.	QUARTZ	
8.	TOPAZ	
9.	CORUNDUM	
10.	DIAMOND	

HARDNESS. The hardness of a mineral is its resistance to abrasion. Hardness can be determined either by trying to scratch a mineral of unknown hardness with a substance of known hardness, or by using the unknown mineral to scratch a substance of known hardness. Hardness is measured on a relative scale called the *Mohs scale of hardness,*[1] which consists of ten common minerals arranged in order of their increasing hardness (table 1.1). In the laboratory, convenient materials other than these ten specific minerals may be used for hardness determination.

In this manual, a mineral that scratches glass will be considered "hard," and one that will not scratch glass will be "soft." In making hardness tests on a glass plate, do not hold the glass in your hands; keep it firmly on the table top. If you think that you have made a scratch on the glass, try to rub the scratch off. What appears to be a scratch may be only some of the mineral that has rubbed off on the glass.

TENACITY. This property is an index of a mineral's resistance to being broken or bent. It is not to be confused with hardness. Some of the terms used to describe tenacity are:

Brittle: The mineral shatters when struck with a hammer, or dropped on a hard surface.

Elastic: The mineral bends without breaking and returns to the original shape when stress is released.

Flexible: The mineral bends without breaking but does not return to its original shape when the stress is released.

SPECIFIC GRAVITY. This is a number that represents the ratio of the weight of the mineral to the weight of an equal volume of water.[2] For purposes of identifying minerals in this manual, it is sufficient to observe the "heft" (specific gravity) of a mineral simply by lifting the hand specimen. Minerals such as graphite (specific gravity = G = 2.2), halite (G = 2.2), and gypsum (G = 2.3) are relatively lightweight when hefted. Quartz (G = 2.6), orthoclase (G = 2.6), and calcite (G = 2.7) are of average specific gravity, whereas garnet (G = 3.5–4.3), corundum (G = 4.0), pyrite (G = 5.0), magnetite (G = 5.2), and galena (G = 7.6) are relatively heavy when hefted. Students may easily be deceived by large specimens that, because of their bulk, are heavy when hefted but that do not possess a high specific gravity.

CRYSTAL FORM. A crystal is a solid bounded by smooth surfaces (crystal faces) that reflect the internal (atomic) structure of the mineral. *Crystal form* refers to the assemblage of faces that constitute the exterior surface of the crystal. *Crystal symmetry* is the geometric relationships between the faces.

Seven crystal systems are recognized by crystallographers, and all crystalline substances crystallize in one of the seven crystal systems (fig. 1.2). Some common substances, such as glass, are often described as crystalline, but in reality they are *amorphous;* that is, they have solidified with no fixed or regular internal atomic structure.

The same mineral always shows the same angular relations between crystal faces, a relationship now recognized as the *constancy of interfacial angles.* The symmetric relationship of crystal faces, related to the constancy of interfacial angles, is the basis for the recognition of the crystal systems.

---

1. Friedrich Mohs (1773–1839) was a German mineralogist and protégé of the famous geologist-mineralogist, Abraham Gottlob Werner (1750–1817) of Freiburg, Germany.

2. A detailed description of the method of determining specific gravity in the laboratory is given in *Minerals and How to Identify Them,* by E. S. Dana, 3d ed., revised by C. S. Hurlbut, Jr. (New York: John Wiley and Sons, Inc., 1963), pp. 75–80.

CRYSTAL SYSTEM	CHARACTERISTICS	EXAMPLES*
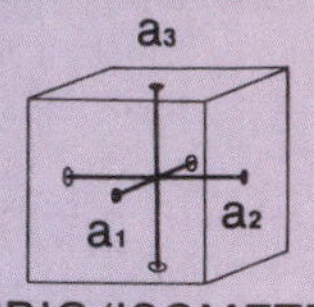 CUBIC (ISOMETRIC)	Three mutually perpendicular axes, all of the same length ($a_1 = a_2 = a_3$). Four-fold axis of symmetry around a_1, a_2, and a_3.	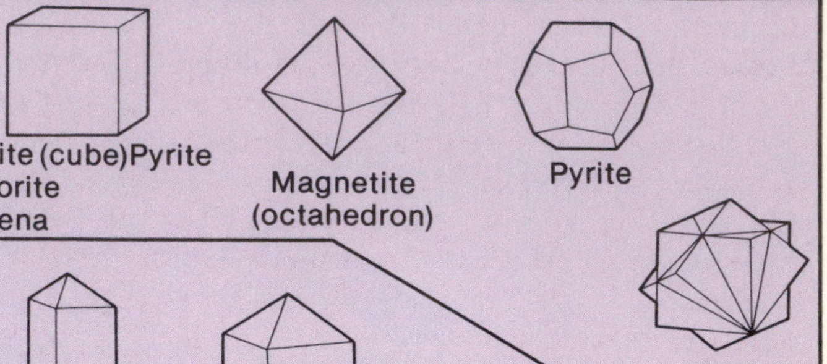 Halite (cube) Pyrite Fluorite Galena • Magnetite (octahedron) • Pyrite • Fluorite (twinned)
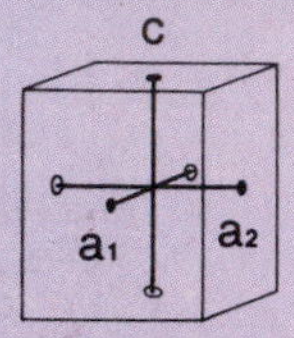 TETRAGONAL	Three mutually perpendicular axes, two of the same length ($a_1 = a_2$) and a third (c) of a length not equal to the other two. Four-fold axis of symmetry around c.	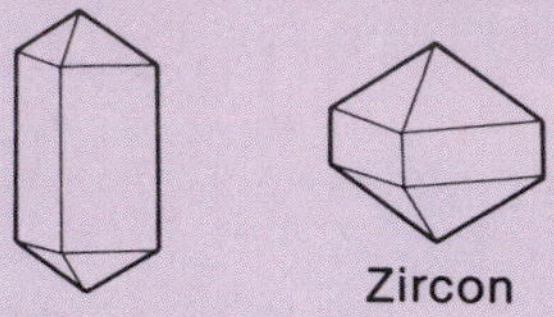 Zircon • Zircon
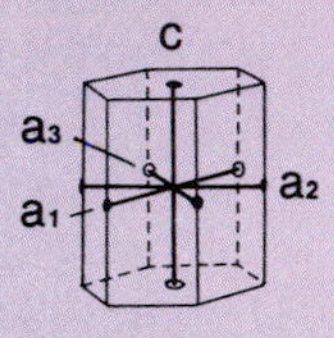 HEXAGONAL	Three horizontal axes of the same length ($a_1 = a_2 = a_3$) and intersecting at 120°. The fourth axis (c) is perpendicular to the other three. Six-fold axis of symmetry around c.	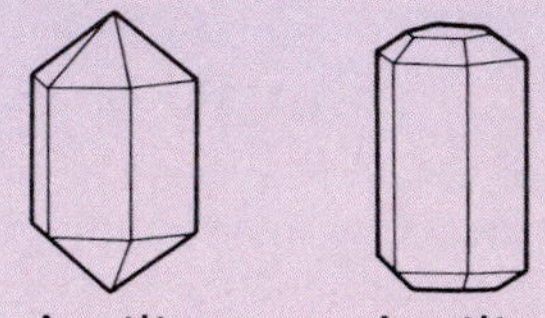 Apatite • Apatite
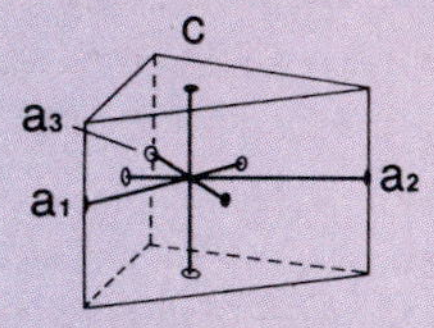 TRIGONAL	Three horizontal axes of the same length ($a_1 = a_2 = a_3$) and intersecting at 120°. The fourth axis (c) is perpendicular to the other three. Three-fold axis of symmetry around c.	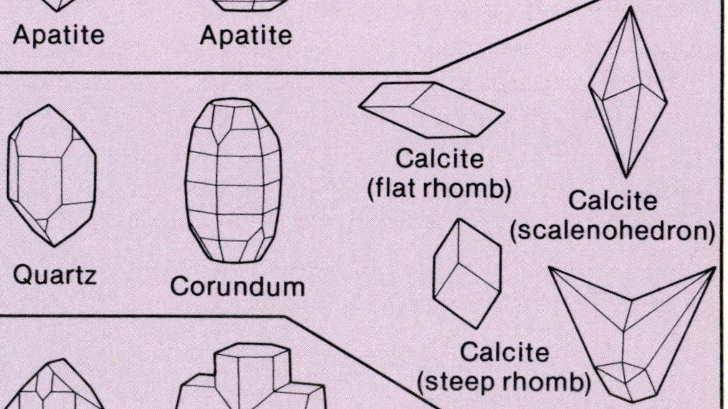 Quartz • Corundum • Calcite (flat rhomb) • Calcite (scalenohedron) • Calcite (steep rhomb) • Calcite (twinned)
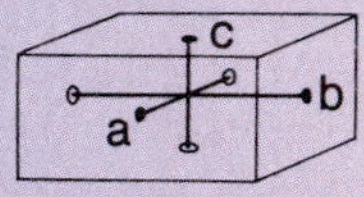 ORTHORHOMBIC	Three mutually perpendicular axes of different length. ($a \neq b \neq c$). Two-fold axis of symmetry around a, b, and c.	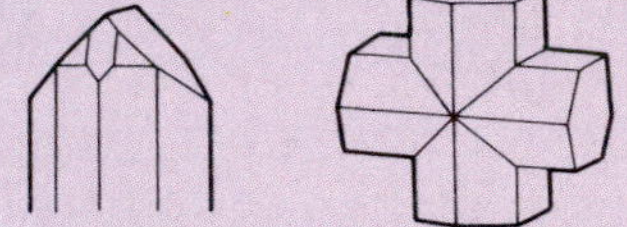 Topaz • Staurolite (twinned)
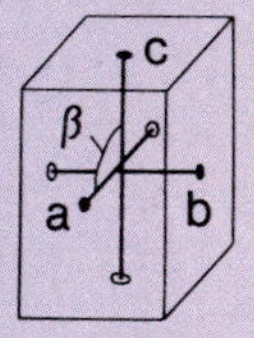 MONOCLINIC	Two mutually perpendicular axes (b and c) of any length. A third axis (a) at an oblique angle (β) to the plane of the other two. Two-fold axis of symmetry around b.	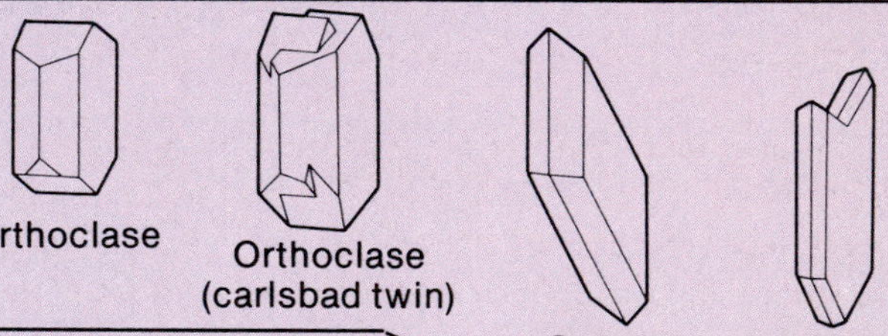 Orthoclase • Orthoclase (carlsbad twin) • Gypsum • Gypsum (twinned)
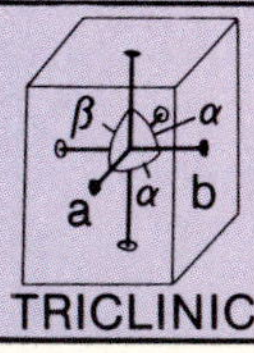 TRICLINIC	Three axes at oblique angles (α, β and α), all of unequal length. No rotational symmetry.	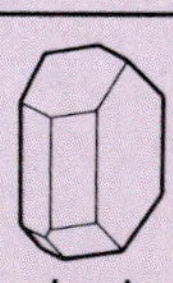 Plagioclase

*Most laboratory collections of minerals for individual student use do not include complete crystals of these minerals. The collection may, however, contain incomplete single crystals, fragments of single crystals, or aggregates of crystals of one or more minerals. The best examples of these and other crystals may be seen on display in most mineralogical museums.

Figure 1.2 Characteristics of the seven crystal systems and some examples. (Copyright © 1974 McGraw-Hill Book Co. (UK) Ltd. From Cox, Price & Harte: *An Introduction to the Practical Study of Crystals, Minerals and Rocks,* Revised Edition. Reproduced by permission.)

Symmetry in a crystal is determined by completing a few geometric operations. For example, a cube has six faces, each at right angles to the adjacent faces. A planar surface that divides the cube into portions such that the faces on one side of the plane are mirror images of the faces on the other side of the plane is called a *plane of symmetry*. A cube has nine such planes of symmetry. In the same way, imagine a line (axis) connecting the center of one face on a cube with the center of the face opposite it. Rotation of the cube about this axis will show that during a complete rotation a crystal face identical with the first face observed will appear in the same position four times. This is a *four-fold axis of symmetry*. Rotation of the cube around an axis connecting opposite corners will show that only three times during a complete rotation does an identical face appear, thus a *three-fold axis of symmetry*.

The seven crystal systems can be recognized by the symmetry they display. In figure 1.2 we have summarized the basic elements of the symmetry for each system and have provided some examples of the *crystal habit* (the crystal form commonly taken by a given mineral) of some minerals you may see in the laboratory or a museum.

Perfect crystals in nature are the exception rather than the rule. They usually form under special conditions where there is open space for them to grow during crystallization. Crystals more commonly are small and distorted. Nevertheless, the internal arrangement of the atoms is fixed, although the external form is not perfectly developed.

Many of the hand specimens you see in the laboratory will be made up of many minute crystals so that few crystal faces, or none, can be seen, and the specimen will appear granular. Other hand specimens may be fragments of larger crystals so that only one or two imperfect crystal faces can be recognized. While perfect crystals are rare, most student laboratory collections contain some reasonably good crystals of quartz, calcite, gypsum, fluorite, or pyrite.

Two or more crystals of some minerals may be grown together in such a way that the individual parts are related through their internal structures. The external form that results is manifested in a *twinned crystal*. Some twins appear to have grown side-by-side (plagioclase), some are reversed or are mirror images (calcite), and others appear to have penetrated one another (fluorite, orthoclase, staurolite). Recognition of twinned crystals may be useful in mineral identification.

A word of caution: Cleavage fragments of minerals such as halite, calcite, and gypsum are often mistaken for crystals. This error is made because the cleavage fragments of these minerals have the same geometric form as the crystal.

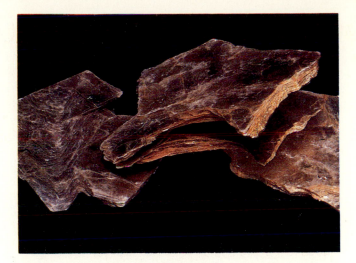

Figure 1.3   Mica is a mineral with excellent basal cleavage. (Photo by Bob Coyle/WCB.)

CLEAVAGE. Cleavage is the tendency of a mineral to break along planes of weakness. The orientation of these planes depends upon the internal structure of the mineral. Some minerals have no cleavage, while others have several. Some minerals exhibit excellent crystal faces but have no cleavage. Quartz is such a mineral.

The cleavage planes of mica (fig. 1.3), calcite, halite, and fluorite are so well-developed that they are detected quite easily. In others, the cleavage surfaces may be so discontinuous as to escape detection by casual inspection. Before deciding that a mineral has no cleavage, turn it around in a good source of light and observe whether there is some position in which the surface of the specimen will "light up" as if it were the reflecting surface of a dull mirror. If so, the mineral has cleavage but the cleavage surface consists of several discontinuous parallel planes minutely separated.

In assigning the number of cleavage planes to a specimen, do not make the mistake of calling two parallel planes bounding opposite sides of a specimen two planes of cleavage. In this case, the specimen has two cleavage surfaces but only one plane of cleavage (i.e., one direction of cleavage). Similarly, a cube of halite has six sides but only three planes of cleavage because the six sides are three parallel pairs of planes.

The angle at which two cleavage planes intersect is diagnostic. This angle can be estimated by inspection. In most cases, you will need to know whether the angle is 90 degrees, almost 90 degrees, or more or less than 90 degrees. The cleavage relationships that you will encounter during the course of your study of common minerals are tabulated for convenience in figure 1.4.

FRACTURE. When a mineral breaks in a direction other than along a cleavage plane, it is said to fracture. Typical kinds of fracture are:

Conchoidal: The surfaces of the fracture are smooth and exhibit fine concentric ridges.

Uneven: The surface of the fracture is irregular and rough. Most minerals have this type of fracture.

Earthy: Soft mineral aggregates, such as kaolinite, possess smooth, but dull, fracture surfaces.

Special Properties

MAGNETISM. The test for magnetism requires the use of a common magnet or magnetized knife blade. Usually, magnetite is the only mineral in your collection that will be attracted by a magnet.

DOUBLE REFRACTION. If an object appears to be double when viewed through a transparent mineral, the mineral is said to have double refraction. Calcite is the best common example.

TASTE. The saline taste of halite is an easy means of identifying the mineral. Few minerals are soluble enough to possess this property. (For obvious sanitary reasons, do not use the taste test on your laboratory hand specimens.)

ODOR. Some minerals give off a characteristic odor when damp. Exhaling on the kaolinite specimen, thus dampening it, causes that mineral to exude a musty or dank odor.

FEEL. The feel of a mineral is the impression gained by handling or rubbing it. Terms used to describe feel are common descriptive adjectives such as soapy, greasy, smooth, rough, and so forth.

CHEMICAL REACTION. Calcite will effervesce (bubble) when treated with cold dilute (0.1N) hydrochloric acid. NOTE: Your laboratory instructor will provide the proper dilute acid if you are to use this test.

Number of Cleavage Planes	Remarks	Examples
1	Usually called *basal* cleavage. Muscovite and biotite are examples.	
2	Two at 90 degrees. Feldspar and pyroxene (augite) have cleavage surfaces that intersect at close to 90 degrees.	
2	Two *not* at 90 degrees. Amphibole (hornblende) cleavage surfaces intersect at angles of about 60 and 120 degrees.	
3	Three at 90 degrees. Mineral with three planes of cleavage that intersect at 90 degrees are said to have *cubic cleavage*. Halite and galena are examples.	
3	Three *not* at 90 degrees. A mineral which breaks into a six-sided prism, with each side having the shape of a parallelogram, has *rhombic cleavage*. Example: calcite.	
4	Four sets of cleavage surfaces in the form of an octahedron produce *octahedral cleavage*. Example: fluorite.	
6	Complex geometric forms. Example: sphalerite.	

Figure 1.4 Descriptive notes on cleavage planes. (Adapted and reprinted with permission from R. D. Dallmeyer, *Physical Geology Laboratory Manual*, Dubuque, Ia. Kendall/Hunt Publishing Company. Copyright © 1978 by Kendall/Hunt Publishing Company.)

Exercise 1. Identification of Common Minerals

Identify the common minerals provided by the instructor. Follow the steps outlined below. Reference may be made to figures 1.5 through 1.20 as an aid to identification. Your laboratory collection may, however, contain some specimens not shown in these plates, and some of the minerals in your collection may appear different from the same minerals shown in the plates, owing to the normal variations within a single mineral species.

1. Select a mineral at random from the group given to you for identification.

2. Determine whether it is light- or dark-colored and whether or not it has a metallic luster. This will place the mineral in either group I, II, or III of table 1.2. (The terms *dark* and *light* are subjective. A mineral that is "dark" to one observer may be "light" to another. This possibility is anticipated in table 1.2 where minerals that could fall in either the "light" or "dark" categories are listed in both groups. The same is true for minerals with metallic and nonmetallic lusters.)

3. *a)* If the mineral falls into either group I or II, proceed to test it, first for hardness and then for cleavage. This will place the mineral with a small group of other minerals in table 1.2. Identification can be completed by noting other diagnostic general or special properties.

   b) If the mineral falls into group III, test it for streak and note other general and special properties such as hardness, color, cleavage, and so on, until the mineral fits the description of one of those given in table 1.2 under group III.

4. To assist you in verifying your identification, refer to the expanded mineral descriptions in table 1.3.

5. When you have identified a mineral by this procedure, write its name on a small piece of paper, place it beneath the mineral identified, and then select another random specimen. After completing all of the specimens assigned to you, ask the instructor to check your identifications. The occurrence and use of each mineral is given in table 1.3. The chemical groupings and composition of some common minerals are presented in table 1.4.

Figure 1.5   Quartz crystal.

Figure 1.6   Rose quartz.

Figure 1.7   Smoky quartz.

Figure 1.8   Chert.

Figure 1.9   Microcline.

Figure 1.10   Plagioclase.

Figure 1.11   Gypsum.

Figure 1.12   Talc.

Figure 1.13   Calcite.

Figure 1.14   Fluorite.

Figure 1.15   Biotite.

Figure 1.16   Olivine.

Figure 1.17   Hematite.

Figure 1.18   Hematite.

Figure 1.19   Limonite.

Figure 1.20   Pyrite.

Table 1.2.   Mineral Identification Key

Abbreviations used: H, hardness; C, cleavage; F, fracture; L, luster; S, streak; and G, specific gravity.

I. Non-metallic, light-colored	Hard, (scratches glass)	Shows cleavage	Flesh-colored; C, planes at nearly right angles; H, 6: G, 2.6. — **ORTHOCLASE**
			Gray to white; C, 2 planes at nearly right angles, cleavage planes show striations; H, 6; G, 2.6-2.75 — **PLAGIOCLASE**
		No cleavage	Colorless or varied colors due to impurities. L, vitreous; hexagonal crystals or massive; H, 7; F, conchoidal; G, 2.65 — **QUARTZ**
			Varied colors due to impurities. L, vitreous to dull; F, conchoidal; G, 2.65. — **CHERT**
			Varied colors due to impurities. Agate commonly banded. L, waxy; H, 7; F, irregular; G, 2.65 — **CHALCEDONY**
			L, vitreous; color, various shades of green, sometimes yellowish; S, white; H, 6.5-7; G, 3.3-3.4. — **OLIVINE**
	Soft, (does not scratch glass)	Shows cleavage	Colorless to white; salty taste; C, cubic; H, 2; G, 2.2. — **HALITE**
			White, yellow, to colorless; H, 3; C, rhombohedral; G, 2.7. — **CALCITE**
			White to transparent; H, 2; occurs as flexible plates; G, 2.3. — **GYPSUM**
			Green to white; soapy feel; H, 1; G, 2.7; S, white. — **TALC**
			Colorless to light yellow; transparent, in thin sheets; S, white; C, 1 excellent plane; occurs as elastic plates; G, 2.75-3.0. — **MUSCOVITE, mica**
			C, fibrous; green to white color; G, 2.6. — **ASBESTOS**
			White, yellow, purple, green; C, octahedral; H, 4; G, 3.2. — **FLUORITE**
		No cleavage	White; earthy odor when damp; F, earthy; H, 2-2.5; G, 2.6. — **KAOLINITE**
			Green to white; soapy feel; H, 1; S, white; G, 2.7. — **TALC**

Modified after F. E. Grout's revised edition of "Kemps Handbook of Rocks," published by D. Van Nostrand Company, N.Y., 1940, page 20.

Table 1.2—*Continued*

		Shows cleavage	Black; H, 5 to 6; C, 2 planes at nearly 90°; G, 3.2–3.6.	AUGITE
			Black; H, 5 to 6; C, 2 planes at about 60°; G, 2.9–3.3.	HORNBLENDE
	Hard, (scratches glass)		Gray to blue-gray; H, 6; 2 planes at nearly right angles; striations on cleavage planes; G, 2.6–2.75.	PLAGIOCLASE
		No cleavage	Hexagonal prisms: triangular pattern of striations on basal faces; H, 9; color is gray, brown, or blue-gray; S, white; G, 4.0.	CORUNDUM
			Red to red-brown; S, white; H, 6.5 to 7.5; fracture resembles a poor cleavage; brittle; G, 3.5–4.3.	GARNET
			L, vitreous; various shades of green, sometimes yellowish; S, white; H, 6.5–7; G, 3.3–3.4	OLIVINE
II. Non-metallic, dark-colored			Gray to gray-black; L, vitreous; H, 7; G, 2.65.	QUARTZ
			Red to brown; L, dull; H, 7; F, conchoidal; G, 2.65.	JASPER, quartz
		Shows cleavage	Brown to black; C, 1 excellent plane; thin plates are elastic; G, 2.7–3.1.	BIOTITE
			Cleavage faces common; L, resinous to submetallic; Yellow-brown color; S, white to pale yellow; H, 3.5–4; G, 3.9–4.1.	SPHALERITE
	Soft, (does not scratch glass)		Dark green to green-black; H, 2 to 2.5; S, white: slippery feel; C, platy; G, 2.7.	CHLORITE
		No cleavage	S, red to red brown; earthy appearance; H, 5–6; G, 5.3.	HEMATITE, soft iron ore
			Green, brown, blue, purple; H, 5; S, white; C, poor basal; G, 3.1.	APATITE
			L, earthy; Yellow, yellow-brown, to brown black color. Apparent H, 1; S, yellow brown; earthy masses.	LIMONITE
			Black; strongly magnetic; H, 6; G, 5.2; S, black.	MAGNETITE
			Lead-pencil black; smudges fingers when handled; H, 1; S, black; G, 2.2.	GRAPHITE
	Black, green-black, or dark green streak.		Pale brass-yellow; H, 6 to 6.5; often occurs as cubes; S, green black to brown black; G, 5.0	PYRITE
III. Metallic Luster			Brass-yellow; tarnishes to purple; H, 3.5 to 4; S, green black; G, 4.2.	CHALCOPYRITE
			Shiny gray; very heavy; C, perfect cubic; H, 2.5; S, lead gray; G, 7.6.	GALENA
	Red streak.		Steel gray color; micaceous appearance; S, red to red-brown; G, 5.3.	HEMATITE
	Yellow, brown, or white streak.		Yellow-brown to dark brown, may be almost black; H, 6; S, yellow brown to brown; G, 3.3–4.3.	LIMONITE
			Cleavage faces common; L. resinous to submetallic; Yellow-brown color; S, white to pale yellow; H, 3.5–4; G, 3.9–4.1.	SPHALERITE

Table 1.3.   Mineral Catalog

APATITE Ca,F Phosphate	Green, brown, blue, purple. Hexagonal. H, 5. C, poor basal. F, conchoidal. L, vitreous to sub-resinous. S, white. G, 3.15–3.2. Massive or granular. Crystals of long prismatic habit. Source of phosphate for fertilizers, safety matches, and munitions.
ASBESTOS Mg,Al Silicate	Various shades of green to white. Monoclinic. H, 2–5, usually 4. F, irregular except in fibrous varieties. L, greasy waxlike in massive varieties, silky when fibrous. S, colorless. G, 2.2–2.65. Occurs in massive, platy, and fibrous forms. Used in manufacturing fireproof materials.
AUGITE (Pyroxene) Ca,Mg,Fe,Al Silicate	Dark green to black. Monoclinic. H, 5–6. C, two planes at nearly right angles. L, vitreous. S, greenish grey. G, 3.2–3.6. Short, stubby, 8-sided prismatic crystals. Often in granular crystalline masses. Common in dark-colored igneous rocks. No commercial value.
BIOTITE K,Mg,Fe,Al Silicate	Dark brown to black. Monoclinic. H, 2.8–3.2. C, perfect basal forming elastic sheets. L, pearly to glassy. S, colorless. G, 2.7. Usually in irregular foliated masses. Accessory mineral in igneous rocks. Associated with muscovite in pegmatites. No commercial value.
CALCITE CaCO$_3$	Colorless, white, yellow. Trigonal. H, 2.5–3. C, perfect rhombohedral. L, vitreous to earthy. S, colorless. G, 2.7. Shows strong double refraction. Effervesces in cold dilute HCl. Chief raw material in Portland Cement. Common in sedimentary rocks.
CHALCEDONY SiO$_2$	Varied colors due to impurities; agate commonly banded. A microcrystalline variety of quartz. H, 7. C, none. F, irregular. L, waxy. G, 2.65. Petrified wood, moss agate, and onyx are other forms. No commercial value.
CHALCOPYRITE CuFeS$_2$	Brass yellow, often tarnished to bronze or purple. Tetragonal. H, 3.5–4. F, uneven. L, metallic. S, greenish black. Brittle. Usually massive. Important ore of copper.
CHERT SiO$_2$	Varied colors due to impurities; usually light in color. A microcrystalline variety of quartz. H, 7. C, none. F, prominent conchoidal. L, vitreous to dull. S, colorless. G, 2.65. No commercial value.
CHLORITE Mg,Fe,Al Silicate	Dark green to greenish black. Monoclinic. H, 2–2.5. C, perfect basal forming flexible but not elastic sheets. L, vitreous to pearly. S, colorless. G, 2.6–2.9. As foliated masses or small flakes. No commercial value.
CORUNDUM Al$_2$O$_3$	Various colors; usually brown, grey, pink, or blue. Trigonal. H, 9. L, adamantine to vitreous. Barrel-shaped crystals, frequently with deep horizontal striations. Gem varieties are ruby and sapphire. Abrasive used in grinding wheels, emery cloth. Ruby and sapphire are gem varieties.
FLINT/JASPER SiO$_2$	Flint: brown to black; jasper: red. Microcrystalline varieties of quartz. H, 7. C, none. F, prominent conchoidal in flint, less prominent in jasper. L, vitreous to dull. S, colorless. G, 2.65.
FLUORITE CaF$_2$	Variable; white, yellow, purple, green. Cubic. H, 4. C, perfect octahedral (four directions parallel to the faces of an octahedron). L, vitreous. S, colorless. G, 3.18. Twins fairly common. Flux in steel-making; glassmaking; source of fluorine for hydrofluoric acid.
GALENA PbS	Lead grey. Cubic. H, 2.5. C, perfect cubic. L, bright metallic. S, lead grey. G, 7.4–7.6. Easily recognized by good cleavage, high specific gravity, and softness. Chief ore of lead.
GARNET Fe,Mg,Ca,Al Silicate	Color varies with composition; most commonly red to red brown but also yellow green. Cubic. H, 6.5–7.5. F, uneven. L, vitreous to resinous. S, white. G, 3.5–4.3. Commonly occurs in crystal form but also in granular masses.
GOETHITE Fe$_2$O$_3$ • nH$_2$O	Yellow, yellow brown, brownish black. Orthorhombic. H, 5–5.5. C, perfect parallel to side pinacoid. L, adamantine to dull. S, yellow brown. Often in radiating forms. Bog iron ore.
GRAPHITE C	Steel grey to black. Hexagonal. H, 1–2. C, perfect basal. L, metallic, sometimes dull earthy. S, black. G, 2.2. Greasy feel. Usually in foliated or scaly masses. "Lead" pencils, electrodes, dry lubricants.
GYPSUM CaSo$_4$ • 2H$_2$O	Colorless, white, grey. Monoclinic. H, 2. C, good in one direction yielding thin sheets. F, conchoidal in one direction, fibrous in another. L, vitreous but also silky and pearly. G, 2.32. Crystals common. Also massive (alabaster) and fibrous. Cements, plasters, and plasterboard.

Table 1.3—*Continued*

HALITE NaCl	Colorless or white; other colors due to impurities. Cubic. H, 2.5. C, perfect cubic. L, transparent to translucent. G, 2.2. Salty taste. Meat packing, domestic salt, pottery glazing.
HEMATITE Fe_2O_3	Steel grey, red grey, to black. Also red to red brown. Trigonal. H, 6, but apparent may be as low as 1. C, basal parting. S, dark red. G, 5.26. Micaceous variety-specular. May occur as crystalline, botryoidal or earthy masses. Most important iron ore mineral.
HORNBLENDE (Amphibole) Ca,Na,Mg,Fe,Al Silicate	Dark green to black. Monoclinic. H, 5–6. C, two planes meeting at angles of 56° and 124°. L, highly vitreous on cleavage faces. S, colorless. G, 2.9–3.2. Long, 6-sided prismatic crystals. May appear fibrous. Coarse to fine-grained crystalline masses also. Common in igneous rocks. No commercial value.
KAOLINITE Al Silicate	White, variously colored by impurities. Monoclinic. H, 2–2.5. F, earthy. L, usually dull earthy. S, colorless. G, 2.6. Usually in claylike masses. Earthy smell when damp. Pottery, brick and tile, porcelain, and other ceramic uses.
LIMONITE $Fe_2O_3 \cdot nH_2O$	Yellow, yellow brown, brown black. Amorphous (not a true mineral). Apparent. H, 1. L, earthy. S, yellow brown. Earthy masses. Yellow ochre paint pigment; minor source of iron (bog iron ore).
MAGNETITE $FeO \cdot Fe_2O_3$	Black. Cubic. H, 6. C, some octahedral parting. L, metallic. S, black. G, 5.2. Strongly magnetic. Usually in granular masses. Important iron ore mineral. Accessory mineral in some igneous rocks.
MUSCOVITE K,Al Silicate	Colorless to light yellow. Monoclinic. H, 2–2.5. C, perfect parallel to the base allowing the mineral to be split into thin elastic sheets. L, vitreous to silky or pearly. S, white. G, 2.75–3. Mostly in thin flakes. Insulating materials, lubricants, wallpaper, paints, fireproofing; occurs in pegmatites (in commercial quantities) and in foliated metamorphics.
OLIVINE $(Mg,Fe)_2SiO_4$	Olive to greyish green, brown. Orthorhombic. H, 6.5–7. F, conchoidal. L, vitreous. S, pale green to white. G, 3.3–3.4. Usually as grains or granular masses. Refractory bricks. Common in peridotite and other ultrabasic igneous rocks.
ORTHOCLASE (K-FELDSPAR) $K(AlSi_3O_8)$	Varies from colorless and white to brown, commonly pinkish-flesh colored. Monoclinic. H, 6. C, two planes at nearly right angles. L, vitreous. S, white. G, 2.6. Distinguished from other feldspars by cleavage and lack of striations. Pottery, enamelware; soft abrasives in scouring powders. Commercial quantities from pegmatites. Common mineral of granites.
PLAGIOCLASE Mixture of Albite, $Na(AlSi_3O_8)$, and Anorthite, $Ca(Al_2Si_2O_8)$	Colorless, grey, blue grey, green, white. Triclinic. H, 6. C, two planes at close to right angles; striations common on good cleavage surfaces. L, vitreous to pearly. G, 2.6–2.75. Well-formed crystals common. Light-colored varieties higher in sodium. Darker varieties higher in calcium. Common in igneous rocks. Minor commerical value as a ceramic.
PYRITE FeS_2	Pale brass yellow. Cubic. H, 6–6.5. F, uneven. L, metallic. S, greenish or brownish black. G, 5.0. Crystals common with striated faces. Also massive. "Fool's Gold." Source of sulfur for sulfuric acid.
QUARTZ SiO_2	Colorless, white; any color may result from impurities. Trigonal. H, 7. C, none. F, conchoidal. L, vitreous. S, none. G, 2.65. Prismatic crystals with striations perpendicular to the long dimension; also a variety of massive forms. Milky quartz, rose quartz, smoky quartz, and amethyst varieties named by color due to impurities. Common in all classes of rocks; glassmaking; crystals used in electronics.
SPHALERITE ZnS	Yellow brown, brown to black. Cubic. H, 3.5–4. C, perfect in six directions at 120°. L, resinous. S, white to reddish brown. G, 3.9–4.1. Usually massive. Cleavage faces common. Important zinc ore.
TALC Mg Silicate	Green to white or silver white. Monoclinic. H, 1.0. C, perfect basal. Massive forms of talc show no visible cleavage. L, pearly to greasy. S, white. G, 2.7–2.8. Soapy feel. Soapstone varieties often dark grey to green. Crystals extremely rare. Paints, ceramics, insecticides, roofing, rubber, paper, toilet articles.

Table 1.4. Chemical Grouping and Composition of Some Common Minerals

| CHEMICAL GROUP | EXAMPLE | |
	MINERAL NAME	CHEMICAL FORMULA*†
ELEMENTS	Native Copper	Cu
	Graphite	C
	Diamond	C
OXIDES	Quartz	SiO_2
	Hematite	Fe_2O_3
	Magnetite	$FeO \cdot Fe_2O_3$
	Limonite	$Fe_2O_3 \cdot nH_2O$
	Corundum	Al_2O_3
SULFIDES	Pyrite	FeS_2
	Chalcopyrite	$CuFeS_2$
	Galena	PbS
	Sphalerite	ZnS
SULFATES	Anhydrite	$CaSO_4$
	Gypsum	$CaSO_4 \cdot 2H_2O$
CARBONATES	Calcite	$CaCO_3$
	Dolomite	$Ca,Mg(CO_3)_2$
PHOSPHATES	Apatite	$Ca_5(PO_4)_3F$
HALIDES	Halite	NaCl
	Fluorite	CaF_2
SILICATES — OLIVINE GROUP	Olivine	$(Mg,Fe)_2SiO_4$
SILICATES — AMPHIBOLE GROUP	Hornblende	Ca,Na,Mg,Fe,Al Silicate
	Asbestos (fibrous serpentine)	Mg,Al Silicate
SILICATES — PYROXENE GROUP	Augite	Ca,Mg,Fe,Al Silicate
SILICATES — MICA GROUP	Muscovite	K,Al Silicate
	Biotite	K,Mg,Fe,Al Silicate
	Chlorite	Mg,Fe,Al Silicate
	Talc	Mg Silicate
	Kaolinite	Al Silicate
SILICATES — FELDSPAR GROUP	Orthoclase (K-Feldspar)	$K(AlSi_3O_8)$
	Plagioclase (Ab,An)	Mixture of Ab and An.
	Albite (Ab)	$Na(AlSi_3O_8)$
	Anorthite (An)	$Ca(Al_2Si_2O_8)$

*Some common elements and their symbols:

Al—Aluminum	Fe—Iron	O—Oxygen
C—Carbon	H—Hydrogen	P—Phosphorous
Ca—Calcium	K—Potassium	Pb—Lead
Cl—Chlorine	Mg—Magnesium	S—Sulfur
Cu—Copper	Mn—Manganese	Si—Silicon
F—Fluorine	Na—Sodium	Zn—Zinc

† Chemical formulas based on Principles of Mineralogy, by W. H. Dennen, New York: Ronald Press, 1960.

187

Rocks

Introduction

A rock is a naturally occurring mass of inorganic or organic material that forms a significant part of the earth's crust. This definition includes soft mud and hard granite, but as used in this manual the term *rock* will be restricted to the "hard" parts of the earth's crust. Most rocks are aggregates of minerals, but some important types, such as coal, are composed mainly of organic materials, with insignificant amounts of true mineral matter.

Rocks are the major units of the earth's crust studied by geologists. Geologists distinguish one rock type from another on the basis of physical properties and mineral composition and deduce the origins of rocks according to an understanding of the various rock-forming processes. In addition, geologists plot the boundaries of different rock masses on a map, a process that leads to a geologic map.

The first part of this section deals with the identification of rocks based on the study of hand specimens. The second part introduces some of the basic principles of the occurrence of rock masses in their natural setting. These principles will be used in understanding the relative ages of adjacent rock masses, and how these relationships can be employed to build a geologic time scale for a given area. The use of simple diagrams and schematic drawings will be used to convey these ideas and concepts.

Three major rock categories are recognized. They are *igneous, sedimentary,* and *metamorphic.* They will be introduced in that order in this manual.

Igneous Rocks

Igneous rocks are aggregates of minerals that crystallize from a *magma.* Magma is a complex high-temperature solution of silicates containing water and various gases that is generated deep beneath the crust and works its way toward the surface. Some magma reaches the earth's surface where it is extruded as *lava,* but other magmas may solidify before they reach the surface.

The reaction series presented in figure 1.21 gives some suggestion as to the formation of the various types of igneous rocks, and it helps to understand why the association of some minerals such as olivine and quartz are rare in nature. The continuous reaction branch indicates that as cooling occurs, there is a continuous reaction between

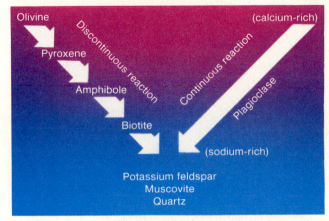

Figure 1.21 Reaction series of rock formation from magma. (From Carla W. Montgomery, *Physical Geology.* Copyright © 1987 Wm. C. Brown Publishers, Dubuque, Iowa. All Rights Reserved. Reprinted by permission.)

the crystals and the melt. The discontinuous branch indicates that as the first crystals form they no longer react with the melt and the composition of the magma changes. The result is that the magma becomes more silicic as cooling occurs, and quartz is the final mineral formed.

Many intrusive bodies have been studied in detail, and the theory presented in this reaction series (commonly called Bowen's Reaction Series after the man who first proposed it) has been verified in principle.

The chemical constituents of a magma determine the kinds of minerals that will be formed as the magma cools and solidifies. The individual mineral grains may be large enough to be identifiable by the naked eye (macroscopic), or they may be so small that they can be distinguished only under a microscope. The *rate of cooling* of the magma determines whether the mineral grains will be microscopic, macroscopic, or a combination of both. The size, shape, and mutual relationships of the minerals in an igneous rock are called the *texture.* The kinds of minerals present in an igneous rock determine its *composition.* These two properties—texture and composition—provide the means for both the identification and the classification of igneous rocks.

15

Textures of Igneous Rocks

The texture of an igneous rock can be described in terms of one of the following:

PHANERITIC TEXTURE. This term applies to an igneous rock in which the constituent minerals are macroscopic in size. The dimensions of the individual crystals range from about 1 mm to more than 5 mm (figs. 1.22–1.25).

APHANITIC TEXTURE. This term is used to describe the texture of an igneous rock composed of crystals that are microscopic in size (fig. 1.26).

PORPHYRITIC TEXTURE. A texture characteristic of an igneous rock in which macroscopic crystals are embedded in a matrix of microscopic crystals (figs. 1.27 and 1.29), or macroscopic crystals of one-size range occur in a matrix of smaller macroscopic crystals. The larger crystals of a rock with a porphyritic texture are called *phenocrysts,* and the matrix in which they are embedded is called the *groundmass.*

VESICULAR TEXTURE. A texture characterized by the presence of *vesicles*—tubular, ovoid, or spherical cavities in the rock (figs. 1.31 and 1.32). The size of the vesicles may be less than 1 mm to several centimeters. Vesicles that are filled with mineral matter are called *amygdules,* and the texture then becomes an *amygdaloidal texture.*

GLASSY TEXTURE. A texture resembling that of glass (fig. 1.33).

Igneous rocks with a phaneritic texture were formed deep beneath the earth's surface where the cooling rate of the parent magma was very slow. Igneous rocks with an aphanitic texture formed at shallow depths or as lava flows. Their cooling rate was more rapid than the rocks with phaneritic textures. A prophyritic texture reflects a two-stage cooling history of a magma. The larger crystals formed first under conditions of slow cooling, but before the magma turned to rock, it migrated to a zone of faster cooling where the remainder of the magma solidified. Vesicular textures are indicative of lava flows in which escaping gases produced the vesicles while the lava was still molten. A glassy texture reflects an extremely rapid rate of cooling in the absence of gases. Lava flows commonly possess a glassy texture.

Colors of Igneous Rocks

The mineral constituents of an igneous rock impart a characteristic color to it. Hence, rock color is used as a first order approximation in establishing the general mineralogic composition of an igneous rock. As already pointed out, "color" is a relative and subjective property when modified only by the adjectives *light, intermediate,* or *dark.* All observers would agree that a white rock is "light-colored," a black rock is "dark-colored," and a rock with half of its constituent minerals white and half of them black is a rock of "intermediate color." Mineral constituents are not all black or white, however. Some are pink, gray, and other colors, a fact that adds to the difficulty encountered in igneous rock classification for the beginning student. Nevertheless, the terms *light, dark,* and *intermediate* are useful terms, especially in the classification of hand specimens with an aphanitic texture. For example, the rocks in figures 1.22 and 1.28 are light-colored. Those in figures 1.24 and 1.29 are intermediate in color, and those in figures 1.25 and 1.26 are dark-colored.

Mineralogic Composition of Igneous Rocks

The minerals of igneous rocks are grouped into two major categories, *primary* and *secondary.* Primary minerals are those that crystallized from the cooling magma. Secondary minerals are those formed after the magma had solidified, and include minerals formed by chemical alteration of the primary minerals, or by the deposition of new minerals in an igneous rock. *Secondary minerals are not important in the classification of igneous rocks.*

Primary minerals consist of two types for the purpose of classifying igneous rocks—*essential minerals* and *accessory minerals.* The essential minerals are those that must be present in order for the rock to be assigned a specific position in the classification scheme (i.e., given a specific name). Accessory minerals are those that may or may not be present in a rock of a given type, but the presence of an accessory mineral in a given rock may affect the name of the rock. For example, the essential minerals of a granite are quartz and K-feldspar (K-feldspar is a common notation for the group of potassium-rich feldspars of which orthoclase is the most common). If a particular granite contains an accessory mineral such as biotite or hornblende, the rock may be called a biotite granite or a hornblende granite, respectively.

The essential minerals contained in the common igneous rocks are quartz, K-feldspar, plagioclase, pyroxene (commonly augite), amphibole (commonly hornblende), and olivine. The key to igneous rock identification is the ability of the observer to recognize the presence or absence of quartz and to distinguish between K-feldspar and plagioclase. Color is of little help in the matter because quartz, K-feldspar, and plagioclase can all occur in the same shade of gray. The distinction between quartz and the feldspars is made by the fact that quartz has no cleavage; macroscopic quartz crystals do not exhibit shiny cleavage faces as the feldspars do.

The distinction between K-feldspar and plagioclase is more difficult because both have cleavage faces that show up as shiny surfaces in phaneritic hand specimens. A pink-colored feldspar is usually K-feldspar (orthoclase), but a white or gray feldspar may be either K-feldspar or plagioclase. Plagioclase, however, has characteristic striations that may be visible on cleavage faces, especially if the crystals are several millimeters in size. Figure 1.10 shows a large fragment of a plagioclase crystal with striations on it. Striations on smaller plagioclase crystals are difficult but not impossible to detect.

Igneous Rock Classification

Table 1.5 is a chart that shows the names of about a dozen common igneous rocks and their corresponding textural and mineralogical compositions. The texture categories are given along the lower left-hand side of the chart, and the chief mineral constituents are shown in the upper part of the chart. *The width of the bar for each mineral is roughly proportionate to the percentage of that particular mineral present. For example, quartz ranges from 0% to about 25% as a mineral constituent, K-feldspar from 0% to 80%, plagioclase from 10% to 70%, olivine from 0% to 90%, amphibole from 0% to 25%, and pyroxene from 0% to 30% (the highest percentage value is equivalent to the widest part of the bar).* These bars are to be used as estimates of percentages for the various mineral constituents, and not as rigorous constraints. This is in keeping with the fact that the percentage of minerals present in a given hand specimen can only be approximated by visual inspection. Moreover, only the minerals in phaneritic rocks and phenocrysts in porphyrys can be identified by visual inspection.

The classification of aphanitic, vesicular, and glassy rocks by macroscopic means is dependent on color alone. For this reason, the chart contains alternate names that can be used if a particular specimen does not fit neatly into one of the boxes on the chart. For example, an aphanitic rock that seems too dark to be a rhyolite but too light to be an andesite can be called a *felsite*. Similarly, a rock intermediate between a granite and a diorite can be labeled a *granitic rock*.

A group of igneous rocks that does not fit easily into the general classification scheme are the *pyroclastic rocks*, those that are accumulations of the material ejected from explosive type volcanoes. The lavas of these volcanoes are characterized by *high viscosity* (they do not flow easily) and high silica content. They are rhyolitic or andesitic in composition. Their mineral constituents are difficult to determine.

The volcanic ash generated from an eruption is known as *tuff* when it becomes consolidated into a rock. Light-colored tuff is called a *rhyolite tuff* (fig. 1.30), and a tuff of intermediate color is called an *andesite tuff*. In some tuffs, small beadlike fragments of volcanic glass occur. These features are called *lapilli* and the rock containing them is a *lapilli tuff*.

A rock composed of the angular fragments from a volcanic eruption is a *volcanic breccia*. A volcanic rock composed of volcanic bombs and other rounded fragments is known as an *agglomerate*.

The classification scheme presented in table 1.5 is a simplified version of a more complex system in which the number of rock names is three to four times the number contained in table 1.5. Simplification at the level of the beginning student is a pedagogical necessity, and if you should encounter a rock in your laboratory collection that does not fit easily into one of the categories shown in table 1.5, it may be because the table has been oversimplified.

Figure 1.22   Granite.

Figure 1.23   Pegmatite.

Figure 1.24   Diorite.

Figure 1.25   Gabbro.

Figure 1.26   Basalt.

Figure 1.27   Basalt porphyry.

Figure 1.28 Rhyolite.

Figure 1.29 Andesite porphyry.

Figure 1.30 Rhyolite tuff.

Figure 1.31 Pumice.

Figure 1.32 Scoria.

Figure 1.33 Obsidian.

Table 1.5. Classification and Identification Chart for Hand Specimens of Common Igneous Rocks

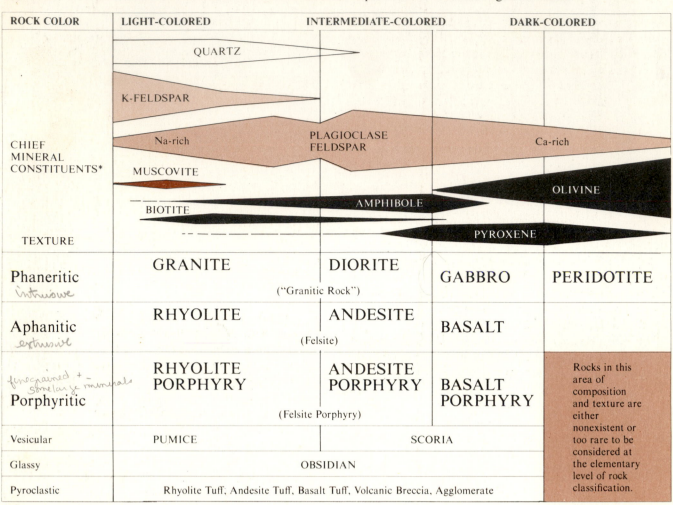

ROCK COLOR	LIGHT-COLORED	INTERMEDIATE-COLORED	DARK-COLORED	
CHIEF MINERAL CONSTITUENTS*	QUARTZ / K-FELDSPAR / PLAGIOCLASE FELDSPAR (Na-rich) / MUSCOVITE / BIOTITE	PLAGIOCLASE FELDSPAR / AMPHIBOLE	PLAGIOCLASE FELDSPAR (Ca-rich) / OLIVINE / PYROXENE	
TEXTURE				
Phaneritic *intrusive*	GRANITE	DIORITE ("Granitic Rock")	GABBRO	PERIDOTITE
Aphanitic *extrusive*	RHYOLITE	ANDESITE (Felsite)	BASALT	
Porphyritic *finegrained + some large minerals*	RHYOLITE PORPHYRY	ANDESITE PORPHYRY (Felsite Porphyry)	BASALT PORPHYRY	Rocks in this area of composition and texture are either nonexistent or too rare to be considered at the elementary level of rock classification.
Vesicular	PUMICE	SCORIA		
Glassy	OBSIDIAN			
Pyroclastic	Rhyolite Tuff, Andesite Tuff, Basalt Tuff, Volcanic Breccia, Agglomerate			

*Muscovite and biotite are accessory minerals and are not essential to the classes of rocks containing them. Amphibole and pyroxene are accessory minerals where shown as a thin dashed line in the granite group.

Exercise 2. Identification of Common Igneous Rocks

A collection of igneous rocks will be provided in the laboratory by your instructor. Spend a little time examining them to see if you can identify textures and individual mineral constituents. Try grouping all the specimens with a phaneritic texture together so as to get some idea of the range of grain sizes that fall into this category.

Using table 1.5, proceed to identify and name each specimen in your collection. Note that you may have more than one specimen of a given rock type. If, for example, you have two or three granites, try to identify the accessory mineral or minerals in each so that you can assign a name that is more definitive than just "granite." For example, a granite containing biotite will be called a *biotite granite*.

When you are reasonably certain that you have identified a particular specimen, write its name on a slip of paper and place the specimen on top. After you have identified all specimens within your collection, the instructor will check those correctly identified and assist you with the ones with which you experienced difficulty.

Sedimentary Rocks

Sedimentary rocks are either derived from preexisting rocks through mechanical or chemical agencies under conditions normal at the surface of the earth, or are composed of accumulations of organic debris. Rock weathering on land produces fragments of rock and mineral detritus that are transported by wind, water, or ice and deposited elsewhere on the earth's surface. Weathering also dissolves rock material and makes it available in solution to streams and rivers that transport it to lakes and oceans, where it may be deposited as a chemical precipitate or evaporite.

After sediment has been deposited, it may be compacted and cemented into a coherent mass or sedimentary rock. The process or processes by which a soft mud or loose deposit of sand is transformed into an indurated rock mass is called *lithification*. Sedimentary rocks exhibit various degrees of lithification. In this manual we will be concerned with sedimentary rocks that are sufficiently cemented or indurated to permit their being displayed and handled as a coherent hand specimen.

Sedimentary rock classification is based on texture and mineralogic composition. Both features are related to the origin of the original sediment, but origin cannot be inferred from a single hand specimen. Therefore, the classification scheme used will emphasize the physical features of the rock rather than its exact mode of origin.

Before describing the system whereby sedimentary rocks are classified, it is necessary to treat briefly sedimentary textures and materials found in sedimentary rocks.

Textures of Sedimentary Rocks

DETRITAL. A detrital (clastic) texture is one that is indicative of an aggregate of mineral grains or rock particles. The size of the individual particles is one of the chief means of distinguishing sedimentary rocks. A single sedimentary rock may contain particles with a wide range of sizes (poorly sorted, figs. 1.36 and 1.38), or it may be composed of particles that fall within a narrow size range (well sorted, figs. 1.34 and 1.37). Usually, but not always, a dominant grain size makes up the major fraction of the rock, and it is the name of this grain size that gives a clastic rock its name. For example, a *sandstone* is a detrital rock composed chiefly of sand-size particles (fig. 1.34).

Detrital particles are named according to their dimensions (i.e., their average grain diameters). Table 1.6 shows the names and size ranges of sedimentary particles according to a classification system called the Wentworth Scale,[3] after the geologist who devised it.

---

3. This classification system of detrital particles is based on the Wentworth Scale after C. K. Wentworth, an American geologist, "A Scale of Grade and Class Terms for Clastic Sediments," *Journal of Geology* 30 (1922): 381.

Table 1.6. The Size Range of Sedimentary Particles

PARTICLE NAME	SIZE RANGE (mm)	APPROXIMATE DECIMAL EQUIV. (mm)
Boulder	greater than 256	256
Cobble	64 – 256	64 – 256
Pebble	4 – 64	4 – 64
Very fine gravel	2 – 4	2 – 4
Very coarse sand	1 – 2	1 – 2
Coarse sand	$\frac{1}{2}$ – 1	.5 – 1
Medium sand	$\frac{1}{4}$ – $\frac{1}{2}$	.25 – .5
Fine sand	$\frac{1}{8}$ – $\frac{1}{4}$	.125 – .25
Very fine sand	$\frac{1}{16}$ – $\frac{1}{8}$	.063 – .125
Coarse silt	$\frac{1}{32}$ – $\frac{1}{16}$	.032 – .063
Medium silt	$\frac{1}{64}$ – $\frac{1}{32}$	.016 – .032
Fine silt	$\frac{1}{256}$ – $\frac{1}{64}$	.004 – .016
Clay	less than $\frac{1}{256}$	.004

Modified from the terminology reported in Lane, E. W. *et al.* 1957. Report of the subcommittee on sediment terminology, *American Geophysical Union Transactions* 2B: 936–38.

Particles larger than about ¼ mm can be distinguished with the naked eye. (Grains of ordinary table salt, for example, range in size from about ½ to ¼ mm. The dot in the letter i is ½ mm in diameter, and a lowercase l is 2½ mm high.) Sand grains smaller than ¼ mm and the larger silt grains can be distinguished with a hand lens; the smaller grains of silt and all clay particles can be distinguished only with a microscope. Sedimentary rocks composed mainly of silt or clay particles are said to have a *dense texture*. The term *dense* is also a general textural term for any rock in which the individual mineral components are microscopic in size (fig. 1.37).

CRYSTALLINE TEXTURE. This texture is characteristic of a sedimentary rock composed of interlocking crystals. If the individual crystals are smaller than ¼ mm, the rock will have a *dense texture* as far as macroscopic examination of a hand specimen is concerned.

AMORPHOUS TEXTURE. This is a very dense texture found in rocks composed of finely divided noncrystalline material deposited by chemical precipitation.

OOLITIC TEXTURE. This texture is formed by spheroidal particles less than 2 mm in diameter, called *oolites*. The oolites are composed of calcium carbonate or silica and are cemented together into a coherent rock mass.

BIOCLASTIC TEXTURE. This texture is produced by the agglomeration of fragments of organic remains, the most common of which are shell fragments (fig. 1.41) and plant fragments. A rock that contains fragments of animal remains (e.g., shells, bones, teeth, leaves) or other recognizable evidence of past life (e.g., footprints, leafprints, worm burrows, etc.) is said to be *fossiliferous* (fig. 1.39). Fossils are commonly embedded in a matrix of sandstone, shale, limestone, or dolomite.

Composition of Sedimentary Rocks

In hand specimens, the mineral constituents of sedimentary rocks generally are less varied than igneous rocks. Because these minerals often occur in microscopic sizes, however, they cannot always be identified easily. Tests for hardness and chemical composition are employed where visual inspection alone fails to identify the primary substance of which a sedimentary rock is composed. The following substances are common in sedimentary rocks.

SILICA (SiO_2). This material occurs as quartz grains, either as the major constituent of the rock or in minor amounts in other rocks. Quartz is the most ubiquitous of all sedimentary minerals. Other forms of silica include *chert,* a dense, noncrystalline form of SiO_2, and *diatomite,* a porous accumulation of the remains of siliceous plants of microscopic size.

CARBONATES. Calcite, $CaCO_3$, and dolomite, $Ca, Mg(CO_3)_2$ are two common sedimentary minerals that occur as the major constituent of limestones and dolomites,[4] or as the cementing material in detrital sediments.

Some calcite occurs in the crystalline form in crystalline limestones. Many shell fragments and most oolites are composed of calcite. Sedimentary rocks containing calcite in any form will effervesce ("fizz") strongly when a drop of *dilute* hydrochloric acid (HCl) is placed on it. Powdered dolomite effervesces weakly with HCl. The names of sedimentary rocks in which a carbonate mineral is present but does not occur as the major constituent are prefixed by the word *calcareous,* which designates the presence of significant amounts of calcite or dolomite. For example, a rock made up of sand-size quartz grains cemented together by calcite would be called a *calcareous quartz sandstone.*

CLAY MINERALS. This term refers to *silicate minerals* that have layered atomic structures. *Clay minerals are not to be confused with clay-size particles that are smaller than 1/256 mm in diameter. All clay minerals occur as clay-size particles, but not all clay-size particles are composed of clay minerals.* Kaolinite is a common clay mineral. Kaolinite is white, but most clay minerals are gray to green and impart a dark color to the sedimentary rock in which they occur. Clay minerals occur most commonly in *shales* and *graywackes,* and many limestones contain appreciable amounts of clay minerals. The adjective used to describe a sedimentary rock with some clay in it is *argillaceous,* as for example, an *argillaceous limestone.*

grey, greenish color

4. Both the mineral and the rock are called *dolomite.* Some geologists use the word *dolostone* for the rock variety of dolomite.

EVAPORITES. Minerals that belong to this group include gypsum, $CaSO_4 \cdot 2H_2O$, and halite, NaCl. Both are formed by chemical precipitation from an aqueous solution.

ROCK FRAGMENTS. Some sedimentary rocks contain very coarse detrital constituents such as pebbles, cobbles, and even boulders. These coarse materials are usually rock fragments rather than single minerals. *Conglomerates* are detrital sedimentary rocks containing rounded cobbles and/or pebbles (fig. 1.36). If the coarse fragments are angular, the rock name applied is *breccia* (fig. 1.38).

FELDSPARS AND OTHER MINERALS. Compared to quartz, calcite, and clay minerals, feldspars do not occur in great abundance in sedimentary rocks. Under certain circumstances, however, some sedimentary rocks may contain significant amounts of feldspar. A feldspar-rich sedimentary rock is called an *arkose* (fig. 1.35). The corresponding adjective is *arkosic.* Feldspars also occur in variable amounts in graywackes.

Other minerals that occur in sedimentary rocks in minor amounts are chlorite, the micas, garnet, hematite, and limonite. The latter two minerals commonly occur as a cement in clastic sediments, and when present even in small amounts, they impart various shades of red, orange, yellow, or brown to the rocks that contain them.

Classification of Sedimentary Rocks

The great variety of sedimentary environments in nature and the gradations between them are responsible for a great diversity of sedimentary rock types. Because of this, a classification scheme encompassing all possible varieties would be unduly complex for the beginning student.

The threefold classification of sedimentary rocks given in table 1.7 is highly simplified and is based on the major constituents of sedimentary rocks: inorganic detrital material, inorganic chemical precipitates, and organic detrital materials (detrital means fragmental). This arrangement is both a logical and a useful guide for an understanding of the similarities and differences in sedimentary rock types encountered by the beginning student of rock classification.

Figure 1.34 Sandstone.

Figure 1.35 Arkose.

Figure 1.36 Conglomerate.

Figure 1.37 Shale.

Figure 1.38 Breccia.

Figure 1.39 Fossiliferous limestone.

Figure 1.40 Chalk.

Figure 1.41 Coquina.

Table 1.7. Classification and Identification Chart for Hand Specimens of Common Sedimentary Rocks

DOMINANT CONSTITUENTS	TEXTURAL FEATURES	COMPOSITION AND/OR DIAGNOSTIC FEATURES	ROCK NAME
INORGANIC DETRITAL MATERIALS	Pebbles and granules imbedded in a matrix of cemented sand grains	Angular rock or mineral fragments.	BRECCIA
		Rounded rock or mineral fragments.	CONGLOMERATE
	Coarse sand and granules	Angular fragments of feldspar mixed with quartz and other mineral grains. Pink feldspar common.	ARKOSE
	Sand-size particles	Rounded to subrounded quartz grains. Color: white, buff, pink, brown, tan.	QUARTZ SANDSTONE
		Calcite and/or dolomite grains. Light colored.	CALCARENITE
	Sand-size particles mixed with clay-size particles	Quartz and other mineral grains mixed with clay. Color: dark gray to gray green.	GRAYWACKE
	Fine-grained, silt and clay-size particles	Mineral constituents not identifiable. Soft enough to be scratched with fingernail. Usually well stratified. Fissile (tendency to separate in thin layers). Color: variable.	SHALE
		Mineral constituents not identifiable. Soft enough to be scratched with fingernail. Massive (earthy). Color: variable.	MUDSTONE
INORGANIC CHEMICAL PRECIPITATES	Dense, crystalline, or oolitic	$CaCo_3$; effervesces freely with dilute HCl. May contain fossils *(fossiliferous)*. Some varieties are *crystalline*. Some varieties are *oolitic*. Color: white, gray, black. Generally lacks stratification.	LIMESTONE
	Dense or crystalline	$Ca,Mg(CO_3)_2$; powder effervesces weakly with dilute HCl. May contain fossils *(fossiliferous)*. Color: variable; commonly similar to limestones. Stratification generally absent in hand specimens.	DOLOMITE
	Dense, porous	$CaCo_3$; effervesces freely with dilute HCl. Color: variable. Contains irregular dark bands.	TRAVERTINE
	Dense (amorphous)	Scratches glass, conchoidal fracture. Color: black, white, gray.	CHERT
	Crystalline	$CaSo_4 \cdot 2H_2O$; commonly can be scratched with fingernail. Color: variable; commonly pink, buff, white.	ROCK GYPSUM
		$NaCl$; salty taste. White to gray. Crystalline. May contain fine-grained impurities in bands or thin layers.	ROCK SALT
ORGANIC DETRITAL MATERIALS	Earthy (bioclastic)	$CaCo_3$; effervesces freely with dilute HCl; easily scratched with fingernail. Microscopic remains of calcareous organisms. White color.	CHALK
		Soft, crumbles, but individual "grains" are harder than glass. Resembles chalk but does not react with dilute HCl. Commonly stratified. Gray to white. Microscopic siliceous plant remains.	DIATOMITE
	Bioclastic	Calcareous shell fragments cemented together.	COQUINA
	Fibrous	Brown plant fibers. Soft, porous, low specific gravity.	PEAT
	Dense	Brownish to brown black. Harder than peat.	LIGNITE
	Dense	Black, dull luster. Smudges fingers when handled.	BITUMINOUS COAL

(handwritten margin notes: "weathering of other Rocks", "precipitation of water solutions", "organics")

24   Part 1

Metamorphic Rocks

Metamorphic rocks are formed when preexisting rocks are changed physically and/or chemically under conditions of high temperature or pressure, or both. These conditions generally prevail deep beneath the earth's surface.

The effects of a magmatic intrusion on the surrounding rock masses (host rocks) result in *contact metamorphism*. Both the heat from the magma and the chemical constituents emanated from it produce mineralogical changes *in situ* in the host rocks. Contact metamorphism is most intense at or near the zone of contact between magma and host rock. The effects of contact metamorphism are progressively diminished in a direction away from the contact zone. Contact metamorphic processes include *recrystallization, chemical recombination,* or *chemical replacement.*

Recrystallization is the process whereby premetamorphic minerals are transformed to larger crystals. The microscopic crystals of calcite in a dense limestone may be recrystallized to interlocking macroscopic crystals in a *marble.*

Chemical recombination is the process whereby the chemical components of two or more minerals in the host rock are recombined to form a new mineral without the addition of any new material. Quartz, SiO_2, and calcite, $CaCo_3$, for example, will recombine under high temperature and pressure to form the mineral wollastonite, $CaSiO_3$.

Chemical replacement, on the other hand, is a process whereby *new* material from the magma is combined with the existing minerals of the host rock to form new minerals. In cases where the host rock has received large amounts of metallic substances from magmatic emanations, minerals of economic value are concentrated into an ore body.

During geologic periods of mountain building, large segments of crustal rocks are deformed. These deformed areas occur in belts hundreds of miles wide and thousands of miles long. Rocks in these belts are subjected to stretching (tensional) and squeezing (compressional) stresses that cause physical changes in the rock. This large scale deformation of rocks is called *regional metamorphism*. Rock masses subjected to stresses under conditions of regional metamorphism are buried at great depth, so they deform or "flow" as a plastic rather than as a brittle solid. This accounts for the fact that many rocks deformed under conditions of regional metamorphism have a texture, called *foliation, that is characterized by a parallel arrangement of platy minerals such as the micas.* Some recrystallization and mineralogical changes are also associated with regional metamorphism.

Both contact and regional metamorphism produce different degrees of change in preexisting rocks. Some rocks may be only slightly metamorphosed if they lie near the outer fringes of the contact zone surrounding an igneous intrusion, or if they lie on the outer margin of a mountain belt that has been subjected to only one period of deformation. Other metamorphic rocks have been so intensely metamorphosed that the texture and mineralogy of the parent rock has been obliterated. For example, the extreme manifestation of regional metamorphism is *granitization,* a process whereby a shale is transformed to a granitelike rock. All gradations between slight and extreme metamorphism occur in nature.

Texture and Composition of Metamorphic Rocks

Metamorphic textures consist of two main types, *foliated* and *nonfoliated.* The mineral constituents of foliated metamorphic rocks are oriented in a parallel or subparallel arrangement. Foliated metamorphic rocks are generally associated with regional metamorphism. The nonfoliated metamorphic rocks exhibit no preferred orientation of mineral grains and most commonly reflect the effects of contact metamorphism.

FOLIATED TEXTURES. Four kinds of foliated textures are recognized, *gneissic* (pronounced nice-ick), *schistose, phyllitic,* and *slaty.*

Gneissic Texture. This is a coarsely foliated texture in which the minerals have been segregated into discontinuous bands, each of which is dominated by one or two minerals. These bands range in thickness from 1 mm to several centimeters. The individual minerals are macroscopic, and impart a "striped" appearance to a hand specimen (figs. 1.43 and 1.44). Light-colored bands commonly contain quartz and feldspar, and the dark bands are commonly composed of hornblende and biotite. A gneissic-textured rock is called a *gneiss.*

Schistose Texture. This is a foliated texture resulting from the parallel to subparallel orientation of platy minerals such as chlorite and mica. Other common minerals present are quartz and amphiboles. A schistose texture lacks the distinct banding of gneissic texture, and the average size of the minerals is generally smaller than those in a gneiss. A rock with a schistose texture is called a *schist.*

Phyllitic Texture. This texture is formed by the parallelism of platy minerals, usually micas, that are barely macroscopic. The parallelism of the micaceous minerals is wavy or crenulated. The preponderance of micaceous minerals imparts a sheen to hand specimens. A rock with a phyllitic texture is called a *phyllite.*

Slaty Texture. This texture is caused by the parallel orientation of microscopic grains. The name for a rock with slaty cleavage is *slate* (fig. 1.42). Slates have a tendency to cleave along parallel planes—a property known as *rock cleavage.* (Rock cleavage is not to be confused with cleavage in a mineral.)

NONFOLIATED TEXTURE. Metamorphic rocks with no visible preferred orientation of mineral grains have a nonfoliated texture. Nonfoliated rocks commonly contain equidimensional grains of a single mineral such as quartz, calcite, or dolomite. Examples of such rocks are *quartzite* (fig. 1.47), which is formed from a quartz sandstone, and *marble* (fig. 1.46), formed from a limestone or dolomite. A conglomerate that has been metamorphosed may retain the original textural characteristics of the parent rock, including the outline and colors of the larger grain sizes such as granules and pebbles. However, because metamorphism has caused recrystallization of the matrix, a metamorphosed conglomerate is called a *meta-conglomerate* (fig. 1.48). These rocks are easily distinguished from their sedimentary equivalents by the fact that they break across the quartz grains, not around them.

A dense textured, nonfoliated rock of metamorphic origin is a *hornfels* (singular and plural). Hornfels have a nondescript appearance because they usually are some

Figure 1.42   Slate, showing characteristic slaty cleavage. (Photo by Bob Coyle/WCB.)

medium to dark shade of gray in color, are lacking in any structural characteristics, and contain few, if any, recognizable minerals in hand specimens.

A commonly overlooked metamorphic rock is *anthracite,* formed by the metamorphism of bituminous coal.

The Naming of Metamorphic Rocks

Foliated metamorphic rocks are named according to their texture (e.g., gneiss, schist, phyllite, or slate). In addition to the root name, the name or names of the dominant or distinctive (but not abundant) mineral(s) may be added as a prefix. Some common examples are: quartz-hornblende gneiss, chlorite schist, biotite-garnet schist, garnet gneiss, amphibole schist, and granite gneiss. The names of slates are commonly modified by a color name as in green slate, black slate, and red slate. In the case of the nonfoliated rocks, a color prefix is also commonplace in the naming of a marble or quartzite. Such names as white marble, pink quartzite, variegated marble (meaning a marble containing streaks of several different colors), are commonplace.

Classification of Metamorphic Rocks

The classification of metamorphic rocks is given in table 1.8. The classification scheme is based on texture and composition. Notice that the basic distinction is between foliated and nonfoliated textures. The rock names given are only the root names because the chart would be too large and cumbersome if all possible varieties of gneisses, schists, marbles, and quartzites were listed. The classification chart can also be used as an identification key.

Figure 1.43   Gneiss.

Figure 1.44   Gneiss.

Figure 1.45   Garnet schist.

Figure 1.46   Pink marble.

Figure 1.47   Quartzite.

Figure 1.48   Metaconglomerate.

Exercise 4. Identification of Common Metamorphic Rocks

A group of metamorphic rock hand specimens will be provided in the laboratory. Use table 1.8 as the classification guide. Because there are only two general categories—foliated and nonfoliated—divide your hand specimens into these two groups. Then proceed with the identification of the foliated specimens, working first with the coarser textured rocks and then with the finer textured ones. Try to use a prefix in naming the specimens of gneiss and schist.

In working with the nonfoliated specimens, remember that marble in all its various forms is composed mainly of calcite or dolomite. Hence, it is softer than glass and will react to dilute HCl in the same way that limestone or dolomite will. Quartzites and meta-conglomerates, on the other hand, are rich in silica (quartz), which is harder than glass.

Your laboratory instructor will check your identifications when you have finished.

Table 1.8. Classification and Identification Chart for Hand Specimens of Common Metamorphic Rocks

TEXTURE	DIAGNOSTIC FEATURES	ROCK NAME
FOLIATED	Macroscopic mineral grains arranged in alternating light and dark bands. Abundant quartz and feldspar in light-colored bands. Dark bands may contain hornblende, augite, garnet, biotite.	GNEISS
	Schistose texture; "coarse" to "fine" grained. Common minerals are chlorite, micas, quartz, garnet, and dark elongate silicate minerals. Feldspar commonly absent.	SCHIST
	Phyllitic texture; aphanitic. Micaceous minerals are dominant. Has "sparkling" appearance.	PHYLLITE
	Dense, microscopic grains. Slaty texture. Color variable. Black and dark gray common. Occurs also in green, dark red, and dark purple.	SLATE
NONFOLIATED	Texture of a conglomerate, but rock breaks across coarse grains as easily as around them. Pebbles may be deformed. Pebbles are commonly granitic, or jasper, chert, quartz, or quartzite.	METACONGLOMERATE
	Crystalline. Hard (scratches glass). Color: white, pink, buff, brown, red, purple.	QUARTZITE
	Dense, dark colored; various shades of gray, gray green, to nearly black.	HORNFELS
	Crystalline. Contains calcite and/or dolomite. Color, variable; fossils in some varieties.	MARBLE
	Black, shiny luster. Conchoidal fracture.	ANTHRACITE COAL

The Occurrence of Rocks

Rock Masses and Geologic Maps

The hand specimens of igneous, sedimentary, and metamorphic rocks that you have just studied originated in the *lithosphere,* the outer skin of the earth. The lithosphere ranges in thickness from 10 to 100 km thick. Some of the rock masses contained in the earth's crust originated at or near the earth's surface on continents, while others formed in oceans or seas that covered continental areas. Still others formed many kilometers beneath the earth's surface in the geologic past, but are now exposed at the surface because the rocks lying above them have been worn away by erosion over long periods of geologic time (many millions of years).

Hand specimens tell us something about the origin of the rock masses from which they were taken. Usually, one can deduce from a hand specimen whether it is igneous, sedimentary, or metamorphic, but because hand specimens are only a very small sample of the rock mass from which they were collected, they reveal very little about the specific geologic circumstances of the parent rock mass.

The occurrence of rock masses and the relationship of one to the other is a matter best studied in the natural setting, or as geologists say, "in the field." Fieldwork involves the tracing of the boundaries or *contacts* of various rock masses as they are revealed in *outcrops,* places at the surface of the earth where the rocks are exposed for visual inspection.

Contacts from many outcrops plotted on a map or aerial photograph result in a *geologic map.* A geologic map shows only a two-dimensional outline of a given rock formation, but by applying a few basic principles it is possible to deduce not only the three-dimensional geometric forms of rock formations, but also to determine their *relative age relationships* with respect to one another.

In Part 4 of this manual we will deal with a more detailed study of geologic maps. At this point, however, we are concerned only with a basic understanding of how rock masses occur in nature, thereby giving more meaning to the hand specimens with which you have already become acquainted.

The Occurrence of Igneous Rocks

Igneous rocks originate from magma deep beneath the surface of the earth. As the magma works its way toward the surface of the earth, some of it congeals into solid crystalline rock, and some magma actually reaches the earth's surface where it becomes lava. Igneous rocks formed at depth are said to be *intruded,* and those that form at the surface are said to be *extruded.*

The rock mass intruded by an igneous rock is referred to as *country rock.* It is not uncommon for pieces of country rock to be engulfed by the invading magma. Fragments of country rock surrounded by igneous rock are called *inclusions.* Country rock can be any kind of rock that existed before an igneous rock was intruded into it. This simple relationship is stated in the general principle that *an igneous rock mass is always younger than the country rock it intrudes.* Another generalization is that *inclusions are older than the igneous rock that encloses them.*

Geometric Forms of Igneous Rock Masses

Several of the more common shapes of igneous rock masses are shown in figure 1.49, and the relationship of rock types to their mode of occurrence is shown in table 1.9. The largest igneous rock mass is the *batholith.* Batholiths are usually phaneritic in texture and granitic in composition. By definition, a batholith crops out over an area of more than 40 square miles. A *stock* is similar in composition to a batholith but, by definition, crops out over an area less than 40 square miles. A *dike* is a tabular igneous intrusive whose contacts cut across the trend of the country rock (fig. 1.50). A *sill* is also tabular in shape but its contacts lie parallel to the trend of the country rock. A *laccolith* is similar to a sill but is generally much thicker, especially near its center where it has caused the country rock to bulge upward.

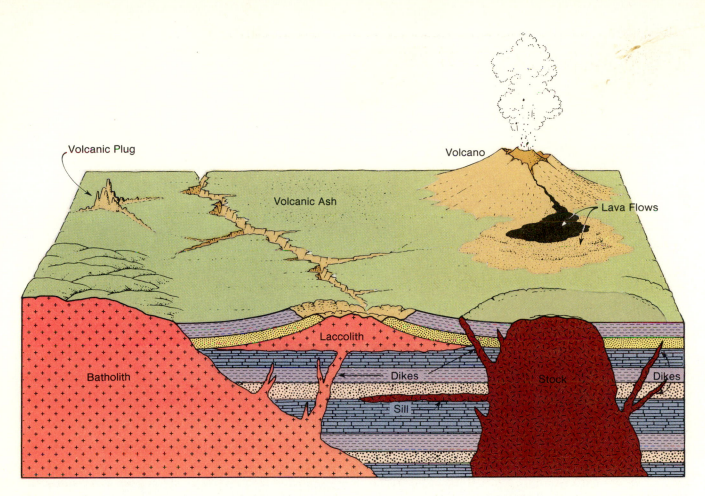

Figure 1.49   Block diagram showing various modes of occurrence of igneous rocks.

Table 1.9.   Relationship of Igneous Rock Types to Their Modes of Occurrence in the Earth's Crust (see table 1.5 for a review of igneous rock types).

	ROCK TYPE	SOME MODES OF OCCURRENCE
EXTRUSIVE	Pumice Scoria	Lava flows, pyroclastics Crusts on lava flows, pyroclastics
EXTRUSIVE	Obsidian	Lava flows
EXTRUSIVE	Rhyolite Andesite Basalt	Lava flows, shallow intrusives
INTRUSIVE	Rhyolite porphyry Andesite porphyry Basalt porphyry	Dikes, sills, laccoliths, intruded at medium to shallow depths
INTRUSIVE	Granite Diorite Gabbro Peridotite	Batholiths and stocks of deep-seated intrusive origin

These five major igneous rock masses—batholiths, stocks, dikes, sills, and laccoliths—can be grouped into two main categories based on the relationship of their contacts to the trend of the enclosing country rock. The contacts of batholiths, stocks, and dikes all cut across the trend of the country rock and, hence, are called *discordant igneous rocks*. Sills and laccoliths, on the other hand, have contacts that are parallel to the trends of the country rock, and are called *concordant igneous rocks*. Table 1.10 summarizes these relationships.

Table 1.10. Relationship between Concordant or Discordant Igneous Bodies and Their Modes of Occurrence

RELATIONSHIP OF IGNEOUS ROCK CONTACT TO COUNTRY ROCK	MODE OF OCCURRENCE
CONCORDANT	SILL, LACCOLITH
DISCORDANT	BATHOLITH, STOCK, DIKE

Figure 1.50   Basaltic dike intruding Precambrian granite north of Lake Superior on the Canadian Shield.

Exercise 5

DUNCAN LAKE AERIAL PHOTOGRAPH, Canada

This photograph (fig. 1.51) was taken from an airplane with a camera whose lens was pointed vertically downward toward the surface of the earth. The picture covers an area of about 3 square miles (2 miles in the east-west direction and 1.5 miles in the north-south direction). The black areas are lakes. Because there are few trees and little vegetation in this part of northern Canada, the rocks are well exposed, or put in geologic terms, there are many outcrops. Fieldwork in the area reveals three rock types, each of which appears as a distinct shade of gray on the photograph. The very light gray areas are granite. The medium gray represents foliated metamorphic rocks, and basalt dikes occur in long narrow outcrop patterns. (When looking at the photo, orient it so that the north arrow is in the upper right-hand corner.)

Figure 1.52A is a schematic vertical section of this area. Geologists call this a *geologic cross-section.* A geologic cross-section shows the mode of occurrence of rocks from the surface of the earth vertically downward.

Using both figures 1.51 and 1.52A, answer the following:

1. Which of the three kinds of rock is the country rock?

2. Does the granite show a discordant or concordant relationship with the country rock? (Examine the small patches of granite in the northeast corner of the photograph for clues. The trend of the country rock in this area is revealed by alternating linear bands of medium gray and somewhat darker gray areas.)

3. What are the modes of occurrence of the granite?

4. Are the basalt dikes younger or older than the granite?

5. List the three rock types by name at the right of the numbered boxes in figure 1.52B with the oldest in box 1 and the youngest in box 3.

NORTH

1200' 0'

Figure 1.51 Aerial photograph of the Duncan Lake area, Northwest Territories, Canada. (Courtesy of the Royal Canadian Air Force.)

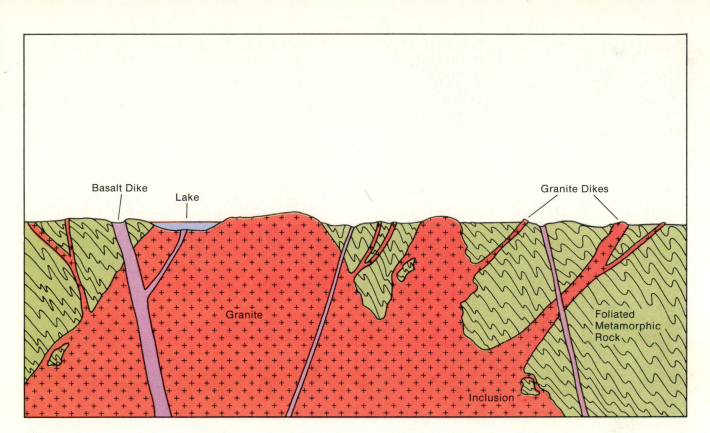

A. Generalized geologic cross-section of Duncan Lake aerial photograph of figure 1.51.

Youngest	3	_____
	2	_____
Oldest	1	_____

B. Geologic column to be completed following instructions in Exercise 5, question 5.

Figure 1.52   Diagrams for use in Exercise 5.

The Occurrence of Sedimentary Rocks

Sedimentary rocks are formed in a wide range of sedimentary environments. Anywhere that sediments accumulate are sites of future sedimentary rocks. *Sedimentary environments* are grouped into three major categories, *continental, mixed continental,* and *marine.* Within each of these broad categories, several subcategories exist. An abbreviated classification of sedimentary environments is given in table 1.11. This table is not intended to show all possible environments of deposition, but rather to illustrate the environments in which the sedimentary rock types listed in table 1.7 might have originated.

Sedimentary environments in a given geographic region do not remain constant throughout geologic time. Areas that are now above sea level were once covered by sea water at various times during the geologic past. Glaciers that existed during the "Ice Age" have since disappeared. As a depositional environment changes with time in a given geographic locality, the kind of sediment changes in response to the new environmental conditions.

Sediments that accumulate in a particular sedimentary environment are generally deposited in layers that are horizontal or almost horizontal. The lateral continuity of these beds or layers reflects the aerial extent and uniformity of the environment in which they were deposited. The thickness of a particular sedimentary layer is a function of the length of time during which the depositional environment remained more or less constant and the rate at which the sediments accumulated in that environment. The physical characteristics (e.g., mineralogic composition, grain size, color, etc.) of a sedimentary rock in an outcrop or hand specimen are collectively known as the *lithology* of the rock.

When the environment of deposition changes in a particular geographic region, the nature of the sediments accumulating there will also change. A bed of sandstone lying on top of a shale bed reflects a change in sedimentary conditions from one in which clay was deposited to one in which sand was deposited. Both the underlying shale and the overlying sandstone represent the passage of geologic time.

Rates of deposition vary widely. A layer of shale only 10 feet thick may represent a longer period of geologic time than a sandstone 100 feet thick. Our concern here is not with the absolute age of sedimentary strata expressed in so many years, but rather with their *relative ages* one to another.

The relative ages of sedimentary strata are determined by the application of two basic geologic principles. The first is *the law of original horizontality, which states that sediments deposited in water are laid down in strata*

Table 1.11. Simplified and Abbreviated Classification of Sedimentary Environments and Some of the Rock Types Produced in Each

SEDIMENTARY ENVIRONMENT	SEDIMENTARY ROCK TYPE
CONTINENTAL	
Desert	SANDSTONE
Glacial	TILLITE
River Beds	SANDSTONE CONGLOMERATE
River Floodplains	SILTSTONE
Alluvial Fans	ARKOSE, CONGLOMERATE, SANDSTONE
Lakes	SHALE, SILTSTONE
Swamps	LIGNITE, COAL
Caves, Hot Springs	TRAVERTINE
MIXED CONTINENTAL AND MARINE	
Littoral (between high and low tide)	SANDSTONE, COQUINA, CONGLOMERATE
Deltaic	SANDSTONE, SILTSTONE, SHALE
MARINE	
Neritic (low tide to edge of shelf)	SANDSTONE, ARKOSE, REEF LIMESTONE CALCARENITE
Bathyl (400 to 4000 m depth)	CHALK, ROCK SALT, ROCK GYPSUM, SHALE, LIMESTONE, GRAYWACKE
Abyssal (depths more than 4000 m)	DIATOMITE, SHALE

that are horizontal or nearly horizontal. The second is the *law of superposition, which states that in any undisturbed sequence of sedimentary rock layers, the layer at the bottom of the sequence is older than the layer at the top of the sequence.*

Sedimentary Strata and Geologic Time

Through the application of the principles used to establish the relative ages of sedimentary strata around the world, a geologic time scale for all of earth history has been pieced together. The geologic time scale as used in North America is shown in table 1.12. The table is arranged with the oldest geologic ages at the bottom and the youngest at the top.

Table 1.12.   Geologic Time Scale as Used in North America. The Chart Shows the Various Geologic Time Units in Order of Their Relative Ages, with the Oldest at the Bottom and the Youngest at the Top. (Absolute ages based on Geological Society of America, The Decade of North American Geology 1983 Time Scale.)

ERA	PERIOD	EPOCH	MAP ABBREV.	COMMON MAP COLOR
CENOZOIC	Quaternary	Holocene	Q	Various shades of gray and yellow
		Pleistocene	Q	
	—1.6 million years—			
	Tertiary	Pliocene	Pl or Tpl	Various shades of orange, yellow orange, and yellow
		Miocene	M or Tm	
		Oligocene	Φ or To	
		Eocene	E or Te	
		Paleocene	Tp	
	—66 million years—			
MESOZOIC	Cretaceous		K	Various shades of green
	Jurassic		J	Various shades of blue green
	Triassic		Ŧ	Various shades of blue
	—245 million years—			
PALEOZOIC	Permian		P or Cpm	Commonly blue, green, purple, pink, lavender purple gray
	Pennsylvanian		ℙ or Cp	
	Mississippian		M or Cm	
	Devonian		D	Various shades of purple pink, lavender, tan brown, red brown, red
	Silurian		S	
	Ordovician		O	
	Cambrian		€	
	—570 million years—			
PRECAMBRIAN			P€ or p€	No standard color

All areas of the earth's surface were not sites of deposition throughout all of geologic time. Some areas, especially mountain regions, contained preexisting rocks from which sedimentary materials were produced by natural decay and physical breakdown. These sediments were in turn transported to sites of deposition by various geologic agents such as running water, glaciers, and wind.

A depositional site may change over geologic time to a site where erosion is taking place. Hence, in any given geographic locality, all of geologic time is not represented by a continuous sequence of strata. Many gaps in the sedimentary record occurred in one locality or another across the face of the earth during the passage of geologic time. Therefore, in order to construct a more complete geologic history of a particular area, fragments of that history recorded in rocks from different localities must be pieced together.

The most thoroughly documented part of the geologic time scale, sometimes referred to as the *geologic column,* comes from the geologic strata deposited during the last 600 million years. This segment of geologic time has been divided into subunits called eras, periods, and epochs. The three major eras in order of decreasing age are the Paleozoic, Mesozoic, and Cenozoic. Each of these is divided into periods, and the periods are further divided into epochs. The names of the eras are based generally on the fossils contained in strata formed during those eras. *A fossil is any evidence of past life such as bones, shells, leaf imprints, and the like.* Paleozoic means "early life,"

Mesozoic means "middle life," and Cenozoic means "recent life." The names of the periods and epochs are based on strata originally studied in Europe during the eighteenth and nineteenth centuries; hence, the names are chiefly European in origin.

The time units of the geologic column are not of equal duration as can be seen from the absolute ages of the time boundaries between the eras in table 1.12. The geologic column as it existed in the early part of the twentieth century was also subdivided by absolute ages, but these dates lacked precision because they were based on inaccurate assumptions. The absolute ages in table 1.12 are based on age determinations on rocks containing radioactive materials that decay at a constant rate. By careful measurement of the components produced by radioactive decay, rocks can be dated with a great deal of precision.

Precambrian rocks are those that were formed prior to the beginning of the Cambrian period. For the sake of simplicity, the Precambrian is not divided into subdivisions in table 1.12. Generally, Precambrian rocks are metamorphic and igneous in type, and the field relationships are extremely complex in many areas where they occur.

The geologic time scale is introduced here primarily to provide a broader context in which the development of simple geologic columns representing very small segments of geologic time can be placed. Other information in table 1.12, such as the abbreviations of geologic time units and

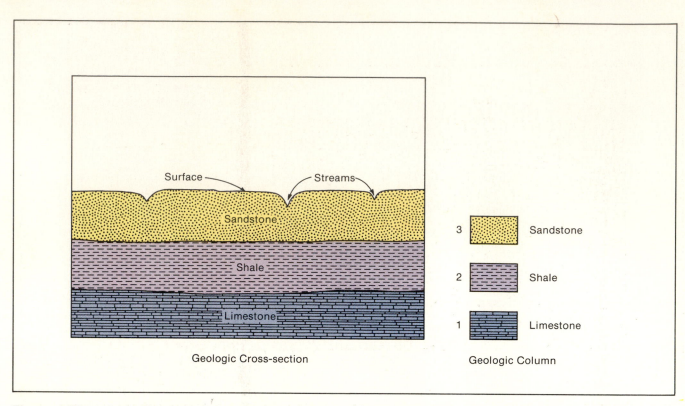

Geologic Cross-section

Geologic Column

3   Sandstone

2   Shale

1   Limestone

Figure 1.53   A simple geologic cross-section and corresponding geologic column showing the relative ages of the three strata with the oldest (1) at the bottom and the youngest (3) at the top. Construction of the geologic column is based on the law of superposition.

standard colors used on geologic maps, is for general information only, and has no application here. However, in Part 4 where geologic maps are considered in detail, the use of these abbreviations and map colors will become apparent.

    We will now consider the means by which a geologic column for a given locality is constructed.

Constructing a Geologic Column

Figure 1.53 shows a series of rock strata in a *geologic cross-section* with a corresponding *geologic column* to the right. The geologic column is constructed by applying the law of superposition to the cross-section. It is apparent that the limestone in figure 1.53 is the oldest, the shale is younger than the limestone, and the sandstone is the youngest of the three formations. The process of sedimentation was continuous during the time it took for these three layers to be deposited; only the depositional environment changed. Figure 1.54 shows the symbols used to portray various rock types on geologic cross-sections and geologic columns.

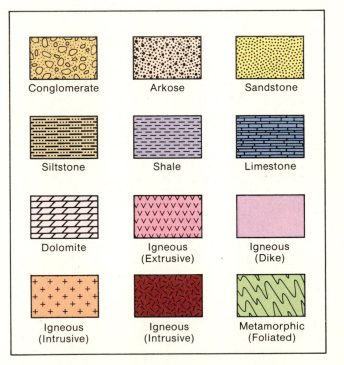

Conglomerate   Arkose   Sandstone

Siltstone   Shale   Limestone

Dolomite   Igneous (Extrusive)   Igneous (Dike)

Igneous (Intrusive)   Igneous (Intrusive)   Metamorphic (Foliated)

Figure 1.54   Symbols used in geologic columns and geologic cross-sections. The colors are used to differentiate rock types only and bear no relationship to the colors used on geologic maps listed in table 1.12.

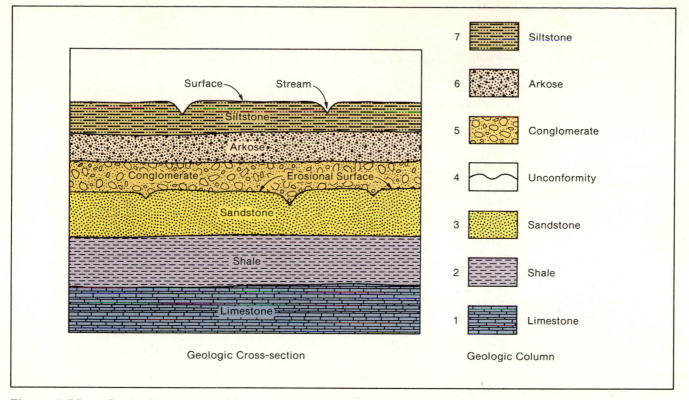

Figure 1.55   Geologic cross-section showing two sequences of sedimentary strata separated by an unconformity. The geologic column at the right shows the sedimentary layers and unconformity of the cross-section arranged in chronological order with the oldest at the bottom and the youngest at the top. The geologic time encompassed by the geologic column cannot be determined because only the relative ages of the rock layers can be deduced from the geologic cross-section.

Unconformities

From the information available in figure 1.53, one can conclude that some time after the sandstone was formed, the marine environment in which it formed changed to an environment of erosion or nondeposition. One cannot deduce from figure 1.53 whether additional strata were layed on top of the sandstone. All that can be said is that a depositional environment gave way to an erosional one some time after the sandstone was formed. Now, if the sea that once covered the area underlain by the three strata in figure 1.53 invaded the area again, and another sequence of strata was deposited, the situation would be depicted as in figure 1.55. The old erosional surface that is now buried beneath the younger strata is called an *unconformity* or a *surface of erosion*. It constitutes an unknown amount of geologic time that elapsed between the cessation of deposition of the older sequence of limestone-shale-sandstone and the upper sequence of conglomerate-arkose-siltstone.

In figure 1.55, the law of superposition still applies in constructing the geologic column. The relative ages of all six formations and the unconformity are shown by the ap-

propriate numbers next to the boxes in the geologic column. The unconformity represents a break of unknown geologic duration in the sedimentary record for this particular locality. It is assumed that this unconformity represents a period of erosion, but the actual length of time in years represented by the unconformity cannot be determined from the information in figure 1.55. All that can be done is to place it in its relative chronological position in the geologic column.

Correlation

In figure 1.55, the two periods of sedimentation and the unconformity between them all represent the passage of geologic time in the same geographic locality. A geologic column constructed therefrom is based only on the rocks that crop out in that particular area. However, by tracing certain formations from this locality to others near by, it may be possible to extend the geologic column to include rocks older or younger.

As an example, consider figure 1.56 in which two geologic cross-sections are shown from two locations separated by a few miles. Locality A contains a sedimentary sequence with six lithologically distinct formations. At locality B, only the upper sedimentary sequence crops out; the lower one is out of sight, presumably beneath the surface and hence, not observable. At locality B, a lava flow lies on top of a siltstone. If, in fact, the siltstone at locality A and the siltstone at B are one and the same formation, they are said to be *correlated*. This being so, it is then possible to construct a geologic column from the stratigraphic information at both localities A and B as shown to the right in figure 1.56. The lava is the youngest formation in the column, and two unconformities are present, one between the siltstone and the lava flow, and one between the conglomerate and the sandstone.

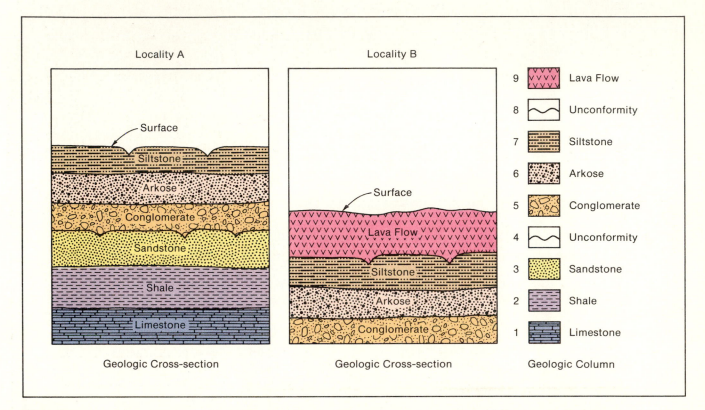

Figure 1.56 The geologic cross-sections shown here are based on two localities, A and B, that are separated by a few miles in which no outcrops exist. By combining the stratigraphy of the two localities, a geologic column using information from both can be constructed. The geologic column is based on the assumption that the conglomerate-arkose-siltstone sequences in both localities are of the same geologic age, in which case they are said to be correlated. Correlation in this case is based on a similar stratigraphic sequence whose constituent beds have lithologic similarities.

Exercise 6. Building a Geologic Column from a Geologic Cross-section

Figure 1.57 is a schematic geologic cross-section based on the mapping of different outcrops from several localities in a geographic area covering about 10 square miles. Formations shown by the same color and symbols are the same age. Study the cross-section and complete the work called for in question 1.

1. Construct a geologic column using the numbered boxes at the right of the cross-section in figure 1.57. Indicate to the right of each box the name of a rock formation (e.g., shale, stock, etc.) or an unconformity in its appropriate position in the column as determined from the adjacent geologic cross-section. Igneous rock masses of the same age should be listed opposite the same box. Remember that the oldest formation is listed to the right of the box labeled "1" in the skeletal geologic column.

2. Fill in each box of the geologic column with an appropriate symbol from figure 1.54. Do not forget unconformities.

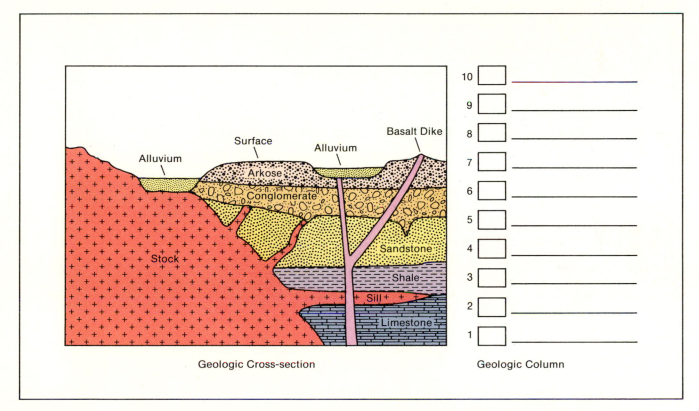

Figure 1.57   Geologic cross-section and uncompleted geologic column. The names of the rock units and any unconformities are to be listed in the appropriate boxes of the geologic column as called for in Exercise 6.

Topographic Maps, Aerial Photographs, and Other Imagery from Remote Sensing

Part 2

Introduction

The planet earth is the laboratory of geologists. They are interested not only in the materials of which the earth is made, but also in the configuration of its surface. Two important tools are used by geologists: maps, and ordinary photographs supplemented by other remote sensing imagery.

A *map* is a representation of part of the earth's surface. Maps that show only the horizontal distribution of earth features and man-made structures are of limited use to the geologist because geologic phenomena are three-dimensional. Therefore, a map that portrays the earth's surface in three dimensions is of particular value to geologists. Maps that meet this requirement are topographic maps. The first section in Part 2 of this manual deals with topographic maps.

Remote sensing is a process whereby the image of a feature is recorded by a camera or other device and reproduced in one form or another as a "picture" of the feature. One of the oldest tools of remote sensing is the camera, which produces images on a photosensitive film. When developed by chemical means, the so-called black-and-white photographs or true-color photographs are the results.

The space age saw the introduction of many other kinds of remote sensing devices that produce new kinds of images such as "radar photographs," false color images, and a variety of other "pictures." These are useful not only to geologists but also to geographers, ecologists, foresters, soil scientists, meteorologists, and the like. The radar picture or image, for example, is made by recording the radiation from earth features in a way that can be resolved into a photolike picture. Radar images from earth-orbiting satellites are used extensively to show cloud cover on televised weather reports and forecasts over most commercial television stations. Radar images made from aircraft are useful in the study of many geologic phenomena.

False color photos or images are made with a device that records infrared radiation from the earth. The resulting image shows earth features in colors that are different from the true colors. For example, vegetations show up as red instead of green on the false color images. False color pictures generally enhance the differences in earth features due to variations in vegetation, soil, water, and rock types.

Photographs or other types of images made by cameras or other sensing devices installed in airplanes, manned spacecraft, or satellites are thus another kind of map that provide geologists with useful tools for analyzing and interpreting the components of a given landscape on earth as well as on the moon and on other planets in the solar system.

The second section of Part 2 introduces the students to aerial photographs and false color images. (A radar map will be introduced in Part 4.) For those interested in pursuing the subject of remote sensing further, attention is directed to the list of references at the end of Part 2.

The goal of Part 2 of this manual is to provide students with a rudimentary knowledge of topographic maps, aerial photographs, and false color images so that they will be able to apply this knowledge in their study of landforms presented in Part 3.

Topographic Maps

A *topographic* map is a graphic representation of the three-dimensional configuration of the earth's surface. Most topographic maps also show land boundaries and other man-made features. The United States Geological Survey (U.S.G.S.), a unit of the Department of the Interior, has been actively engaged in the making of a series of standard topographic maps of the United States and its possessions since 1882.

The features shown on topographic maps may be divided into three groups: (1) *relief,* which includes hills, valleys, mountains, plains, and the like; (2) *water features,* including lakes, ponds, rivers, canals, swamps, and the like; and (3) *culture,* works of man such as roads, railroads, land boundaries, and similar features (fig. 2.1). Relief is printed in brown; water in blue; and culture, including geographical names, in black; some recent maps now show major roads in red. (See fig. 2.2 for some standard map symbols.) On some maps, forests are printed in green.

Standard topographic maps of the U.S.G.S. cover a *quadrangle* of area that is bounded by *parallels of latitude* (forming the northern and southern margins of the map) and by *meridians of longitude* (forming the eastern and western margins of the map). The published maps have different *scales.* A map scale is a means of showing the relationship between the size of an object or feature indicated on a map and the corresponding actual size of the same object or feature on the ground.

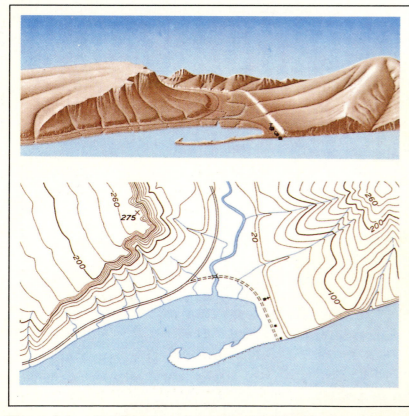

THE USE OF SYMBOLS IN MAPPING

These illustrations show how various features are depicted on a topographic map. The upper illustration is a perspective view of a river valley and the adjoining hills. The river flows into a bay that is partly enclosed by a hooked sandbar. On either side of the valley are terraces through which streams have cut gullies. The hill on the right has a smoothly eroded form and gradual slopes, whereas the one on the left rises abruptly in a sharp precipice from which it slopes gently, and forms an inclined table land traversed by a few shallow gullies. A road provides access to a church and two houses situated across the river from a highway that follows the seacoast and curves up the river valley.

The lower illustration shows the same features represented by symbols on a topographic map. The contour interval (the vertical distance between adjacent contours) is 20 feet.

Figure 2.1   Landforms as shown on a topographic map. Contour lines are printed in brown, water features in blue, and man-made structures in black. (Courtesy of the U.S.G.S.)

Figure 2.2   Standard symbols used on ⟶ topographic maps published by the U.S.G.S.

Topographic Map Symbols

BOUNDARIES

National
State or territorial
County or equivalent
Civil township or equivalent
Incorporated-city or equivalent
Park, reservation, or monument
Small park

LAND SURVEY SYSTEMS

U.S. Public Land Survey System:
  Township or range line
    Location doubtful
  Section line
    Location doubtful
  Found section corner; found closing corner
  Witness corner; meander corner

Other land surveys:
  Township or range line
  Section line
  Land grant or mining claim; monument
  Fence line

ROADS AND RELATED FEATURES

Primary highway
Secondary highway
Light duty road
Unimproved road
Trail
Dual highway
Dual highway with median strip
Road under construction
Underpass; overpass
Bridge
Drawbridge
Tunnel

BUILDINGS AND RELATED FEATURES

Dwelling or place of employment: small; large
School; church
Barn, warehouse, etc.: small; large
House omission tint
Racetrack
Airport
Landing strip
Well (other than water); windmill
Water tank: small; large
Other tank: small; large
Covered reservoir
Gaging station
Landmark object
Campground; picnic area
Cemetery: small; large

RAILROADS AND RELATED FEATURES

Standard gauge single track; station
Standard gauge multiple track
Abandoned
Under construction
Narrow gauge single track
Narrow gauge multiple track
Railroad in street
Juxtaposition
Roundhouse and turntable

TRANSMISSION LINES AND PIPELINES

Power transmission line: pole; tower
Telephone or telegraph line
Aboveground oil or gas pipeline
Underground oil or gas pipeline

CONTOURS

Topographic:
  Intermediate
  Index
  Supplementary
  Depression
  Cut; fill

Bathymetric:
  Intermediate
  Index
  Primary
  Index Primary
  Supplementary

MINES AND CAVES

Quarry or open pit mine
Gravel, sand, clay, or borrow pit
Mine tunnel or cave entrance
Prospect; mine shaft
Mine dump
Tailings

SURFACE FEATURES

Levee
Sand or mud area, dunes, or shifting sand
Intricate surface area
Gravel beach or glacial moraine
Tailings pond

VEGETATION

Woods
Scrub
Orchard
Vineyard
Mangrove

COASTAL FEATURES

Foreshore flat
Rock or coral reef
Rock bare or awash
Group of rocks bare or awash
Exposed wreck
Depth curve; sounding
Breakwater, pier, jetty, or wharf
Seawall

BATHYMETRIC FEATURES

Area exposed at mean low tide; sounding datum
Channel
Offshore oil or gas: well; platform
Sunken rock

RIVERS, LAKES, AND CANALS

Intermittent stream
Intermittent river
Disappearing stream
Perennial stream
Perennial river
Small falls; small rapids
Large falls; large rapids
Masonry dam
Dam with lock
Dam carrying road
Intermittent lake or pond
Dry lake
Narrow wash
Wide wash
Canal, flume, or aqueduct with lock
Elevated aqueduct, flume, or conduit
Aqueduct tunnel
Water well; spring or seep

GLACIERS AND PERMANENT SNOWFIELDS

Contours and limits
Form lines

SUBMERGED AREAS AND BOGS

Marsh or swamp
Submerged marsh or swamp
Wooded marsh or swamp
Submerged wooded marsh or swamp
Rice field
Land subject to inundation

Map series and quadrangles

Each map in a U. S. Geological Survey series conforms to established specifications for size, scale, content, and symbolization. Except for maps which are formatted on a County or State basis, USGS quadrangle series maps cover areas bounded by parallels of latitude and meridians of longitude.

Map scale

Map scale is the relationship between distance on a map and the corresponding distance on the ground. Scale is expressed as a ratio, such as 1:25,000, and shown graphically by bar scales marked in feet and miles or in meters and kilometers.

Standard edition maps

Standard edition topographic maps are produced at 1:20,000 scale (Puerto Rico) and 1:24,000 or 1:25,000 scale (conterminous United States and Hawaii) in either 7.5 x 7.5- or 7.5 x 15-minute format. In Alaska, standard edition maps are available at 1:63,360 scale in 7.5 x 20 to 36-minute quadrangles. Generally, distances and elevations on 1:24,000-scale maps are given in conventional units: miles and feet, and on 1:25,000-scale maps in metric units: kilometers and meters.

The shape of the Earth's surface, portrayed by contours, is the distinctive characteristic of topographic maps. Contours are imaginary lines which follow the land surface or the ocean bottom at a constant elevation above or below sea level. The contour interval is the elevation difference between adjacent contour lines. The contour interval is chosen on the basis of the map scale and on the local relief. A small contour interval is used for flat areas; larger intervals are used for mountainous terrain. In very flat areas, the contour interval may not show sufficient surface detail and supplementary contours at less than the regular interval are used.

The use of color helps to distinguish kinds of features:

Black – cultural features such as roads and buildings.
Blue – hydrographic features such as lakes and rivers.
Brown – hypsographic features shown by contour lines.
Green – woodland cover, scrub, orchards, and vineyards.
Red – important roads and public land survey system.
Purple – features added from aerial photographs during map revision. The changes are not field checked.

Some quadrangles are mapped by a combination of orthophotographic images and map symbols. Orthophotographs are derived from aerial photographs by removing image displacements due to camera tilt and terrain relief variations. An orthophotoquad is a standard quadrangle format map on which an orthophotograph is combined with a grid, a few place names, and highway route numbers. An orthophotomap is a standard quadrangle format map on which a color enhanced orthophotograph is combined with the normal cartographic detail of a standard edition topographic map.

Provisional edition maps

Provisional edition maps are produced at 1:24,000 or 1:25,000 scale (1:63,360 for Alaskan 15-minute maps) in conventional or metric units and in either a 7.5 x 7.5- or 7.5 x 15-minute format. Map content generally is the same as for standard edition 1:24,000- or 1:25,000-scale quadrangle maps. However, modified symbolism and production procedures are used to speed up the completion of U.S. large-scale topographic map coverage.

The maps reflect a provisional rather than a finished appearance. For most map features and type, the original manuscripts which are prepared when the map is compiled from aerial photographs, including hand lettering, serve as the final copy for printing. Typeset lettering is applied only for features which are designated by an approved name. The number of names and descriptive labels shown on provisional maps is less than that shown on standard edition maps. For example, church, school, road, and railroad names are omitted.

Provisional edition maps are sold and distributed under the same procedures that apply to standard edition maps. At some future time, provisional maps will be updated and reissued as standard edition topographic maps.

National Mapping Program indexes

Indexes for each State, Puerto Rico, the U. S. Virgin Islands, Guam, American Samoa, and Antarctica are available. Separate indexes are available for 1:100,000-scale quadrangle and county maps; USGS/Defense Mapping Agency 15-minute (1:50,000-scale) maps; U. S. small scale maps (1:250,000, 1:1,000,000, 1:2,000,000 scale; State base maps; and U. S. maps); land use/land cover products; and digital cartographic products.

Series	Scale	1 inch represents approximately	1 centimeter represents	Size (latitude x longitude)	Area (square miles)
Puerto Rico 7.5-minute	1:20,000	1,667 feet	200 meters	7.5 x 7.5 min.	71
7.5-minute	1:24,000	2,000 feet (exact)	240 meters	7.5 x 7.5 min.	49 to 70
7.5-minute	1:25,000	2,083 feet	250 meters	7.5 x 7.5 min.	49 to 70
7.5 x 15-minute	1:25,000	2,083 feet	250 meters	7.5 x 15 min.	98 to 140
USGS/DMA 15-minute	1:50,000	4,166 feet	500 meters	15 x 15 min.	197 to 282
15-minute	1:62,500	1 mile	625 meters	15 x 15 min.	197 to 282
Alaska 1:63,360	1:63,360	1 mile (exact)	633.6 meters	15 x 20 to 36 min.	207 to 281
County 1:50,000	1:50,000	4,166 feet	500 meters	County area	Varies
County 1:100,000	1:100,000	1.6 miles	1 kilometer	County area	Varies
30 x 60-minute	1:100,000	1.6 miles	1 kilometer	30 x 60 min.	1,568 to 2,240
U. S. 1:250,000	1:250,000	4 miles	2.5 kilometers	1° x 2° or 3°	4,580 to 8,669
State maps	1:500,000	8 miles	5 kilometers	State area	Varies
U. S. 1:1,000,000	1:1,000,000	16 miles	10 kilometers	4° x 6°	73,734 to 102,759
U. S. Sectional	1:2,000,000	32 miles	20 kilometers	State groups	Varies
Antarctica 1:250,000	1:250,000	4 miles	2.5 kilometers	1° x 3° to 15°	4,089 to 8,336
Antarctica 1:500,000	1:500,000	8 miles	5 kilometers	2° x 7.5°	28,174 to 30,462

How to order maps

Mail orders. Order by map name, State, and series/scale. Payment by money order or check payable to the U. S. Geological Survey must accompany your order. Your complete address, including ZIP code, is required.

Maps of areas *east* of the Mississippi River, including Minnesota, Puerto Rico, the Virgin Islands of the United States, and Antarctica.

Maps of areas *west* of the Mississippi River, including Alaska, Hawaii, Louisiana, American Samoa, and Guam.

Eastern Distribution Branch
U. S. Geological Survey
1200 South Eads Street
Arlington, VA 22202

Western Distribution Branch
U. S. Geological Survey
Box 25286, Federal Center
Denver, CO 80225

The geologist makes use of topographic maps because they provide him with a means whereby he can observe earth features in *three dimensions*. Unlike other maps, topographic maps show natural features to a fair degree of accuracy in terms of length, width, and vertical height or depth. Thus by examination of a topographic map and through an understanding of the symbols shown thereon, the geologist is able to interpret earth features and draw conclusions as to their origin in the light of geologic processes.

In the United States, Canada, and other English-speaking countries of the world, most maps produced to date have used the English system of measurement. That is to say, distances are measured in feet, yards, or miles; elevations are shown in feet; and water depths are recorded in feet or fathoms (1 fathom = 6 feet). In 1977, in accordance with national policy, the U.S.G.S. formally announced its intent to convert all of its maps to the metric system. As resources and circumstances permit, new maps published by the U.S.G.S. will show distances in kilometers and elevations in meters. The conversion from English to metric units will take many decades in the United States. In this manual, most maps used will be those published by the U.S.G.S. *prior* to the adoption of the metric system, because the metric maps are still insufficient in number to portray the great diversity of geologic features presented in this manual. However, some examples of map scales in metric units will be given to acquaint students with the system.

To help students familiarize themselves with the metric system and its relationship to the English system, table 2.1 is provided as a convenient reference.

Table 2.1.   Units of Measurement in the English and Metric Systems, and the Means of Converting from One to the Other

A. English Units of Linear Measurement
   12 inches = 1 foot
   3 feet = 1 yard
   1 mile = 1,760 yards, 5,280 feet, 63,360 inches

B. Metric Units of Linear Measurement
   10 millimeters = 1 centimeter
   100 centimeters = 1 meter
   1,000 meters = 1 kilometer

C. Conversion of English Units to Metric Units

symbol	when you know	multiply by	to find	symbol
in.	inches	2.54	centimeters	cm
ft.	feet	30.48	centimeters	cm
yd.	yards	0.91	meters	m
mi.	miles	1.61	kilometers	km

D. Conversion of Metric Units to English Units

symbol	when you know	multiply by	to find	symbol
mm	millimeters	0.04	inches	in.
cm	centimeters	0.4	inches	in.
m	meters	3.28	feet	ft.
m	meters	1.09	yards	yd.
km	kilometers	0.62	miles	mi.

Elements of a Topographic Map

Map Scale

Three scales are commonly used in conjunction with topographic maps: (1) *fractional,* (2) *graphic,* and (3) *verbal.*

1. A *fractional scale* is a fixed ratio between linear measurements on the map and corresponding distances on the ground. It is sometimes called the *representative fraction* or R. F.

$$\text{Example: R. F. } 1{:}62{,}500 \text{ or } \frac{1}{62{,}500}.$$

This notation simply means that 1 unit on the map equals 62,500 of the *same units* on the ground. Thus, a line 1 inch long on the map represents a horizontal distance of 62,500 inches long on the ground. (Note that the numerator of the R. F. is always 1.)

2. A *graphic scale* is simply a line or bar drawn on the map and divided into units that represent ground distances.

3. A *verbal scale* is a convenient way of stating the relationship of map distance to ground distance. For example, "1 inch equals 1 mile" is a verbal scale and means that 1 inch on the map equals 1 mile on the ground. Or, "1 cm equals 1 km" means that 1 centimeter on the map equals 1 kilometer on the ground.

Converting from One Scale to Another

It is sometimes necessary to convert from a fractional scale to a verbal or graphic scale, or from verbal to fractional. This involves simple problems in arithmetic. The following are examples of some conversion problems.

Example 1. Convert an R. F. of 1:125,000 to a verbal scale in terms of inches per mile. In other words, how many miles on the ground are represented by 1 inch on the map?

Solution: It is best to express the R. F. as an equation in terms of what it actually means. Thus,

1 unit on the map = 125,000 units on the ground.

We see from this basic equation that we are dealing in the same units on the map and on the ground. Since we are interested in *1 inch* on the map, let us substitute *inches* for *units* in the above equation. It then reads:

1 inch on the map = 125,000 inches on the ground.

The left side of the equation is now complete, since we initially wanted to know the ground distance represented by *1 inch* on the map. We have yet to resolve the right side of the equation into miles because the problem specifically called for a ground distance in miles. Since there are 5,280 feet in a mile and 12 inches in a foot,

$$5,280 \times 12 = 63,360 \text{ inches in one mile.}$$

If we divide 125,000 inches by the number of inches in a mile, we arrive at:

$$\frac{125,000}{63,360} = 1.97 \text{ miles.}$$

The answer to our problem is one inch = 1.97 miles or, for all practical purposes, 2 miles.

Example 2. On a certain map, 2.8 inches is equal to 1.06 miles on the ground. Express this verbal scale as an R. F.

Solution: Again, it is wise to follow simple steps to avoid confusion. We know that, by definition,

$$\text{R. F.} = \frac{\text{distance on the map}}{\text{distance on the ground}}.$$

Since we were given a map distance of 2.8 inches and the equivalent ground distance of 1.06 miles, it is a simple matter to substitute these in our basic equation. Thus,

$$\text{R. F.} = \frac{2.8 \text{ inches on the map}}{1.06 \text{ miles on the ground}}.$$

It is now necessary to change the denominator of the equation to inches because both terms in the R. F. must be expressed in the same units. Since there are 63,360 inches in 1 mile, we must multiply the denominator by that number:

$$1.06 \times 63,360 = 67,161.6 \text{ inches.}$$

Now our equation reads,

$$\text{R. F.} = \frac{2.8 \text{ inches}}{67,161.6 \text{ inches}}.$$

To complete the problem, we need to express the numerator as unity, and so the numerator must be divided by itself. In order not to change the value of the fraction, we must also divide the denominator by the same number, 2.8. The problem resolves itself into,

$$\text{R. F.} = \frac{2.8 \div 2.8}{67,161.6 \div 2.8} = \frac{1}{23,986}$$

or rounded off to $\frac{1}{24,000}$.

If you can follow these two examples and understand them completely, you can handle any type of conversion problem.

Exercise 7. Problems in Scale Conversion

1. Convert the following representative fractions to ground distances equal to one inch on a map. (Fill in the following blanks.)
   a) 1:24,000   1 inch = _____ feet
   b) 1:31,680   1 inch = _____ miles
   c) 1:48,000   1 inch = _____ miles
   d) 1:62,500   1 inch = _____ miles
   e) 1:250,000   1 inch = _____ miles
   f) 1:1,000,000 1 inch = _____ miles
   g) 1:1,000,000 1 cm = _____ km

2. A map of unknown scale shows two TV transmitting towers. On the map the towers are 1.2 inches apart and the actual ground distance between them is 1,000 feet. What is the R. F. of the map?

3. A straight stretch of road on an aerial photograph was found to be 500 yards long. The same road segment measured on the photograph was three-fourths of an inch. What is the R. F. of the photograph?

4. A foreign map has an R. F. of 1:500,000. How many kilometers on the ground are represented by 10 centimeters on the map?

5. The bar scales shown in figure 2.3 appear on one of the new U.S.G.S. maps. Derive the R. F. of the map first using the kilometer scale and then using the mile scale. Which of the two systems, English or metric, is simpler to handle in the derivation of the R. F.? Why?

Figure 2.3   Two bar scales from a recent U.S.G.S. topographic map.

Map Coordinates and Land Divisions

The earth's surface is arbitrarily divided into a system of reference coordinates called *latitude* and *longitude*. This coordinate system consists of imaginary lines on the earth's surface called *parallels* and *meridians* (fig. 2.4). Both of these are best described by assuming the earth to be represented by a globe with an axis of rotation passing through the North and South poles. Meridians are circles drawn on this globe that pass through the two poles. Meridians are labeled according to their positions, in degrees, from the zero meridian, which by international agreement passes through Greenwich near London, England. The zero meridian is commonly referred to as the *Greenwich* or *Prime meridian*. If meridional lines are drawn for each 10 degrees in an easterly direction from Greenwich (toward Asia) and in a westerly direction from Greenwich (toward North America), a family of great circles will be created. Each one of the great circles is labeled according to the number of degrees it lies east or west of the Greenwich or zero meridian. The 180° west meridian and the 180° east meridian are one and the same great circle, and constitute the International Date Line.

A great circle represents the intersection of a plane that connects two points on the surface of a sphere and passes through the center of the sphere, in this case the earth. The intersection of the plane with the surface divides the earth into two equal halves—hemispheres—and the arc of the great circle is the shortest distance between two points on the spherical earth.

Another great circle passing around the earth midpoint between the two poles is the equator. It divides the earth into the Northern and Southern Hemispheres. A family of lines drawn on the globe parallel to the equator constitute the second set of reference lines needed to locate a point on the earth accurately. These lines form circles that are called *parallels of latitude,* and are labeled according to their distances in degrees north or south of the equator. The parallel that lies halfway between the equator and the North Pole is latitude 45° North, and the North Pole itself lies at latitude 90° North.

This system of meridians and parallels thus provides a means of accurately designating the location of any point on the globe. Santa Monica, California, for example, lies at about longitude 118° 29' West and latitude 34°01' North. For increased accuracy in locating a point, degrees may be subdivided into 60 subdivisions known as *minutes* indicated by the notation '. Minutes may be subdivided into 60 subdivisions known as *seconds* indicated by the notation ". Thus, a position description might read 64° 32' 32" East, 44° 16' 18" South.

Meridional lines converge toward the North or South Poles from the equator, and the length of a degree of longitude varies from 69.17 statute miles at the equator to zero miles at the poles. Latitudinal lines, on the other hand, are always parallel to each other. However, because the earth is not a perfect sphere but is slightly bulged at the

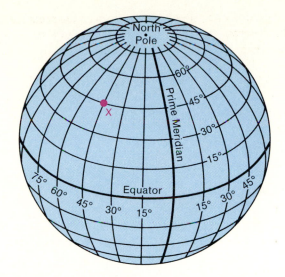

Figure 2.4 Generalized system of meridians and parallels. The location of point X is longitude 45° West, latitude 45° North.

equator, a degree of latitude varies from 68.7 statute miles at the equator to 69.4 statute miles at the poles. Thus, the area bounded by parallels and meridians is not a true rectangle. U.S.G.S. quadrangle maps are also bounded by meridians and parallels, but on the scale at which they are drawn, the convergence of the meridional lines is so slight that the maps appear to be true rectangles. The U.S.G.S. standard quadrangle maps embrace an area bounded by 7½ minutes of longitude and 7½ minutes of latitude. These quadrangle maps are called 7½-minute quadrangles. Other maps published by the U.S.G.S. are 15-minute quadrangles and a few of the older ones are 30-minute quadrangles.

Meridians always lie in a true north-south direction, and parallels always lie in a true east-west direction. *Magnetic north,* however, is the direction toward which the north-seeking end of a magnetic compass needle will point. Because the magnetic poles are not coincident with the north and south ends of the earth's rotational axis, magnetic north is different from true north except on the meridian that passes through the magnetic North Pole. The angle between true north and magnetic north is called the *declination,* and is normally shown on the lower margin of most U.S.G.S. maps for the benefit of those who use a compass in the field to plot geological or other data on a base map (e.g., a U.S.G.S. standard quadrangle map).

Range and Township

For purposes of locating property lines and land descriptions on legal documents, another system of coordinates is used in the United States and some parts of Canada. This system is tied into the latitude and longitude coordinate system, but functions independently of it. The basic block of this system is the *section,* a rectangular block of land one mile long and one mile wide. An area containing

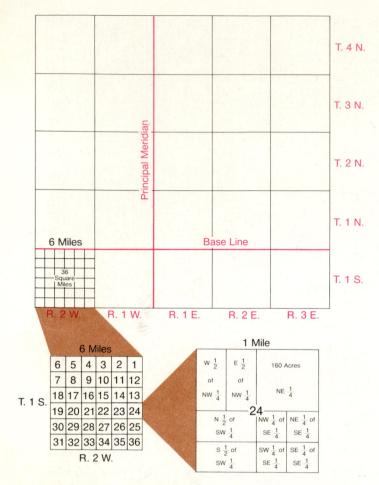

Figure 2.5 Standard land divisions used in the United States and some parts of Canada.

Each township consists of 36 square-mile sections that are numbered according to the system shown in figure 2.5. One section of land contains 640 acres.

For purposes of locating either man-made or natural features in a given section, an additional convention is employed. This consists of dividing the section into quarters called the northeast one-quarter (NE¼), the southwest one-quarter (SW¼), and so on. Sections may also be divided into halves such as the north half (N½) or west half (W½). The quarter sections are further divided into four more quarters or two halves, depending on how refined one wants to make the description of a feature on the ground. For example, an exact description of the 40 acres of land in the extreme southeast corner of the map of section 24 in figure 2.5 would be as follows: SE¼ of the SE¼ of Section 24, T. 1 S., R. 2 W. This style of notation will be used throughout the manual in referring the student to a particular feature on a map used in an exercise.

Figure 2.6 is an aerial photograph of rural Iowa. The rectangular network of roads tends to follow section lines, and the cultivated fields generally conform in shape and size according to the system of land divisions shown in figure 2.5.

Because the maps used in this manual are only *selected parts* of standard U.S.G.S. quadrangle maps, the township and range lines numbering system, magnetic declination, and other data usually found on the lower margin of the maps may not be included. Where necessary, these data will be supplied in the text of the exercise.

Topography

Topography is the configuration of the land surface and is shown by means of *contour lines* (fig. 2.1). A contour line is an imaginary line on the surface of the earth connecting points of equal elevation. The *contour interval* (C.I.) is the difference in elevation of any two adjacent contour lines. Elevations are given in feet or meters above mean sea level. The shore of a lake is, in effect, a contour line because every point on it is at the same level (elevation).

Contour lines are brown on the standard U.S.G.S. maps. The C.I. is usually constant for a given map and may range from 5 feet for flat terrain to 50 or 100 feet for a mountainous region. C.I.'s for U.S.G.S. maps using the metric system are 1, 2, 5, 10, 20, 50, or 100 meters, depending upon the smoothness or ruggedness of the terrain to be depicted. Usually, every fifth contour line is printed in a heavier line than the others and bears the elevation of the contour above sea level. In addition to contour lines, the heights of many points on the map, such as road intersections, summits of hills, and lake levels are shown to the nearest foot or meter on the map. These are called *spot elevations* and are accurate to within the nearest foot or meter. More precisely located and more accurate in elevation are *bench marks,* points marked by brass plates

36 sections is called a *township.* While it was the intent of the original government land surveyors to make each section an exact square of land, many sections and townships are irregular in shape because of surveying errors and other discrepancies in laying out the network.

The north-south lines marking township boundaries are called *range lines,* and the east-west boundaries are called *township lines.* The coordinate system of numbering townships has as a reference or beginning point, the intersection of a meridian of longitude and a parallel of latitude, called *principal meridian* and *baseline,* respectively. A particular township is identified by stating its position north or south of the baseline and east or west of the principal meridian. The system of numbering township and range lines is shown in figure 2.5. The letter *T* along the right-hand margin of the large map stands for the word *township,* and the letter *R* stands for the word *range.* The notation T. 3 N., R. 1 E. is read, "Township three north, Range one east." Under this system, each township has a unique numerical designation. Several principal meridians and baselines are used in the coterminous United States, so that the township and range coordinate numbers are never very large.

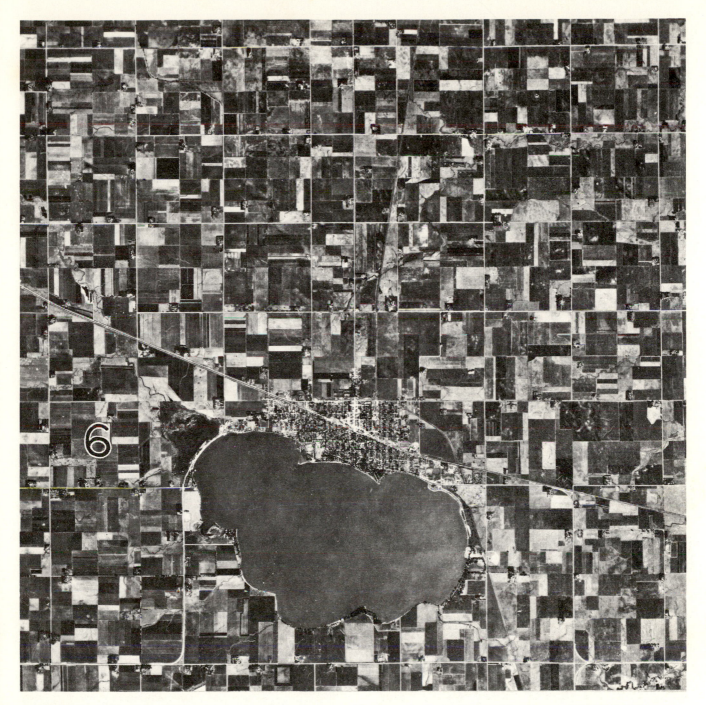

Figure 2.6 Aerial photograph of a rural area in Iowa. The town is Storm Lake. (Courtesy of the U.S.G.S. Photo was taken on September 23, 1950.)

permanently fixed on the ground, and by crosses and elevations, preceded by the letters *BM,* printed in black on the map.

Following are several general statements regarding contour lines.

1. When contour lines cross streams, they bend upstream; that is, the segment of the contour line near the stream forms a "V" with the apex pointing in an upstream direction.
2. Contours do not intersect or cross unless they become merged in a vertical or overhanging cliff.
3. Closed contours appearing on the map as ellipses or circles represent hills or knobs.

4. Closed contours with hachures, short lines pointing toward the center of the closure (downslope), represent closed depressions. The outer hachured contour line has the same elevation as the lower adjacent regular contour line.
5. Steep slopes are shown by closely spaced contours, gentle slopes by widely spaced contours.
6. The difference in elevation between the highest and the lowest point of a given area is the *maximum relief* of that area.

Examine figure 2.1 for the relationship of contour lines to topography.

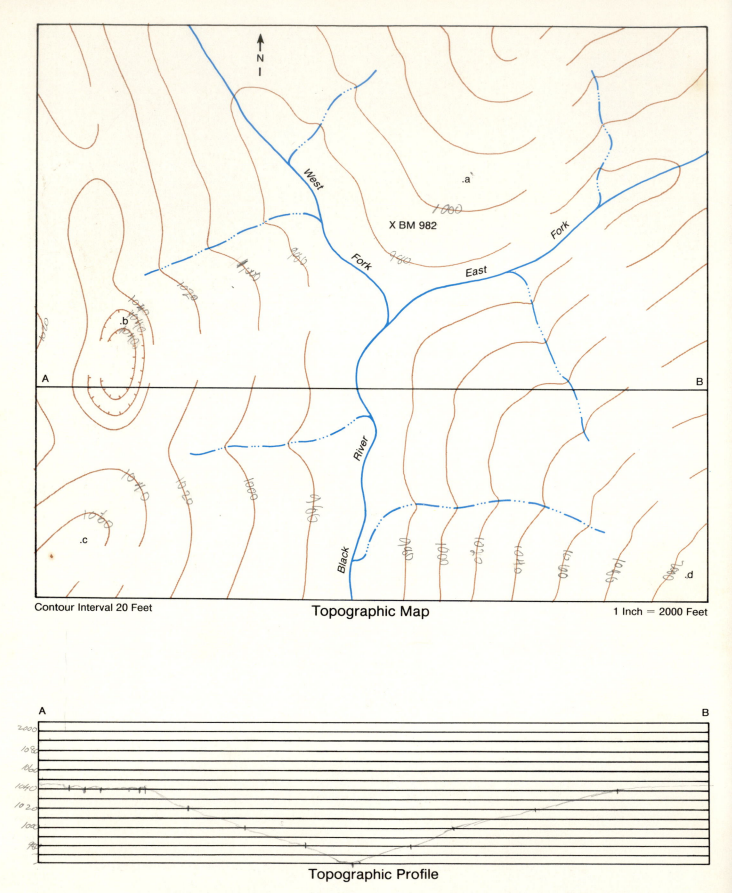

Contour Interval 20 Feet

Topographic Map

1 Inch = 2000 Feet

Topographic Profile

Figure 2.7 Topographic map and topographic profile to be used in connection with Exercise 8, question 1, and Exercise 9, question 1.

Exercise 8. Problems on Contour Lines

1. Study the map in the upper part of figure 2.7. Note the elevation of the bench mark (B.M. 982). This notation means that the point marked X is 982 feet above sea level. Label each contour line with its proper elevation and determine the approximate elevation of points labeled *a, b, c,* and *d*. (Note that the C.I. is 20 feet.) The elevation of each contour line should be written in the space between the broken contour lines (see fig. 2.1 for example).

2. The map shown in figure 2.8 shows spot elevations, drainage lines, and a lake. Using a C.I. of 5 feet, construct a topographic map of the area. (Note: All of the spot elevations will not lie on one of the lines you draw. To draw the contour lines, you will have to interpolate where the lines should be drawn between points of known elevation. This requires some judgement as to the placement of the lines. For example, only one spot elevation on the map is shown at the 65-foot elevation. To draw the 65-foot contour line you must estimate where other points at that elevation lie relative to known points of elevation. Thus, the 65-foot contour would probably be drawn closer to a spot elevation of 64 feet than it would to a spot elevation of 70 feet.)

3. What is the R. F. of the map in figure 2.7?

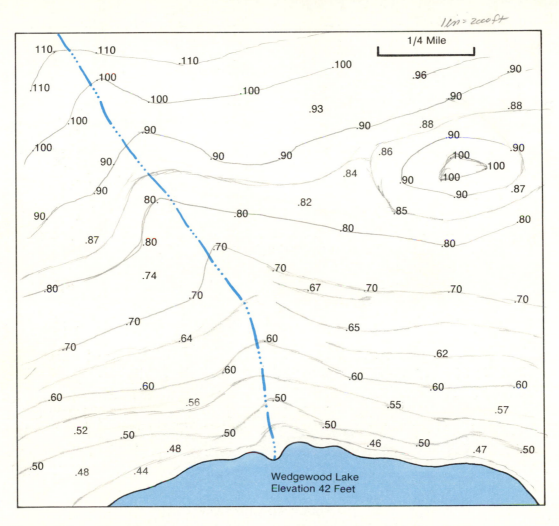

Figure 2.8   Map showing drainage lines and spot elevations to be used with Exercise 8, question 2.

Topographic Profiles

A *topographic profile* is a diagram that shows the change in elevation of the land surface along any given line. It represents graphically the "skyline" as viewed from a distance. Features shown in profile are viewed along a horizontal line of sight, whereas features shown on a map or in *plan view* are viewed along a vertical line of sight. Topographic profiles can be constructed from a topographic map along any given line.

The *vertical scale* of a profile is arbitrarily selected and is usually, but not always, larger than the horizontal scale of the map from which the profile is drawn. Only when the horizontal and vertical scales are the same is the profile *true,* but in order to facilitate the drawing of the profile and to emphasize differences in relief, a larger vertical scale is used. Such profiles are *exaggerated profiles*.

Instructions for Drawing a Topographic Profile

Figure 2.9 shows the relationship of a topographic profile to a topographic map. It should be examined in connection with the following instructions:

1. The line along which a cross-section is to be constructed may be defined by an actual line drawn on the map, or by two points on the map that determine the terminals of the line of cross-section.
2. Examine the line along which the profile is to be drawn and note the difference between the highest and lowest contours crossed by it. The difference between them is the maximum relief of the profile. Cross-sectional paper divided into 0.1 inch or 2 mm squares makes a good base on which to draw a profile. Use a vertical scale as small as possible so as to keep the amount of vertical exaggeration to a minimum. For example, for a profile along which the maximum relief is less than 100 feet, a vertical scale of 0.1 inch or 2 mm = 5, 10, 20, or 25 feet is appropriate. For a profile with 100 to 500 feet of maximum relief, a vertical scale of 0.1 inch or 2 mm = 40 or 50 feet is proper. If the maximum relief along the profile is between 500 and 1,000 feet, a vertical scale of 0.1 inch or 2 mm = 80 or 100 feet is adequate. When the maximum relief is greater than 1,000 feet, a vertical scale of 0.1 inch or 2 mm = 200 feet is appropriate. The general rule for guidance in the selection of a vertical scale is: the greater the maximum relief, the smaller the scale. Label the horizontal lines of the profile grid with appropriate elevations from the contours crossed by the line of the profile. Every other line on a 0.1 inch or 2 mm grid is sufficient.
3. Place the edge of the cross-sectioned paper along the line of profile. Opposite each intersection of a contour line with the line of profile, mark a short dash at the edge of the cross-sectioned paper. If the contour lines are closely spaced, only the heavy or *index* contours need to be marked. Also mark the positions of streams, lakes, hilltops, and significant cultural features on the line of profile. At the edge of the paper, label the elevation of each dash.
4. Drop these elevations perpendicularly to the corresponding elevations represented by the horizontal lines on the cross-sectional paper.
5. Connect these points by a smooth line and label significant features such as streams and summits of hills. Add the horizontal scale and write a title on the profile.

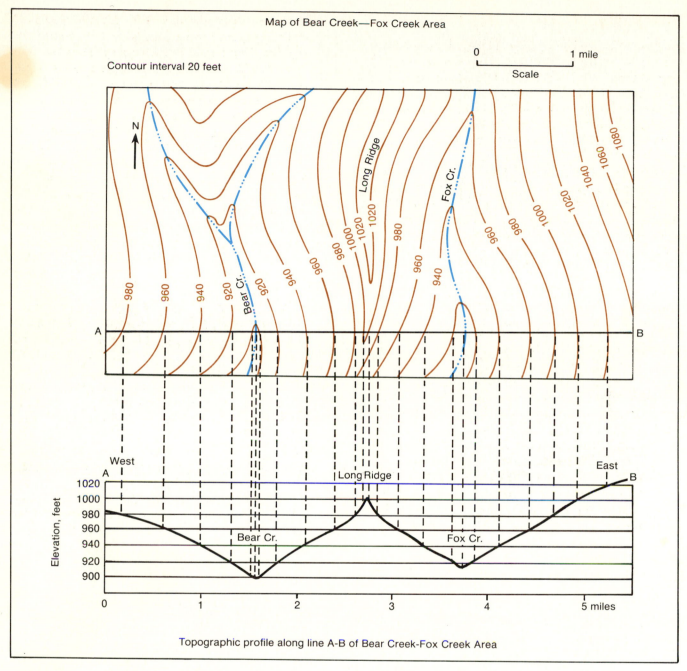

Figure 2.9 A topographic profile drawn along line A–B on the map of the hypothetical Bear Creek—Fox Creek area. See text for step-by-step instructions.

Exercise 9. Drawing Profiles from Topographic Maps

1. Draw a profile along line A–B of the topographic map in figure 2.7.

2. Refer to the Delaware Map of figure 2.11. Three north-south red lines and three east-west red lines intersect at approximately one-mile intervals. These north-south and east-west lines define the boundaries of *sections*. Each section is numbered, and the number is printed in red in the center of each section.

   On the grid of figure 2.10, draw a north-south profile (in pencil) along the red line that defines the western boundary of sections 10, 3, and 34. The beginning point of the profile is the southwest corner of section 10, and the ending point is the shore of Lake Superior, which has an elevation of 602 feet above sea level. (Part of Lake Superior is indicated along the northern part of the Delaware Map, fig. 2.11.) The vertical scale has been estab-

lished as 0.1 inch = 40 feet. The horizontal scale is the same as the map scale. Significant features along the line of profile have already been labeled.

3. What is the horizontal scale of the profile in feet per inch?

4. What is the vertical scale of the profile you have drawn in feet per inch?

5. By what factor is the vertical scale exaggerated?

6. In order to visualize more clearly the effect of vertical exaggeration, redraw the profile using a vertical scale of 0.1 inch = 80 feet. Use the grid in figure 2.10 and label the horizontal lines (according to the new scale) on the right-hand margin of the grid. The horizontal line now labeled 600 feet will be labeled 800, and the line now labeled 800 will be 1,200 on the new scale.

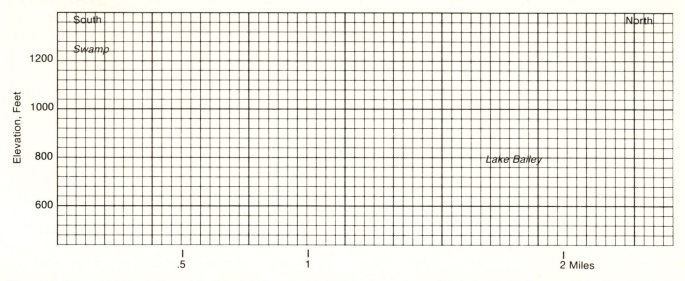

Figure 2.10   Grid to be used in drawing a north-south topographic profile from the Delaware Map (fig. 2.11). The south end of the profile is the southwest corner of section 10, and the north end is the shore of Lake Superior. The line of profile is coincident with the western boundaries of sections 3, 10, and 24.

Exercise 10. Topographic Map Reading

The following questions are based on the Delaware Map, figure 2.11.

1. What is the R. F. of the map?

2. What is the C.I. of the map?

3. If the C.I. was twice what it actually is, would there be a greater or lesser number of contour lines?

4. What is the map distance in feet between the highest point on Mt. Lookout (1,335 feet) and the nearest point on the south shore of Lake Bailey?

5. If you walked along the line described in question 4, would the actual distance walked be greater, less, or the same as the map distance? Explain.

6. What is the maximum relief of the map area?

7. Give the location of the following features using the township, range, section, etc. (the northernmost east-west red line separates T. 58 N. from T. 59 N., and the entire map area falls in R. 30 W.).
   a) North Pond.
   b) Confluence of Bailey Creek with Silver River.

8. Refer to figure 2.5. Notice that each square mile of land (1 section) is comprised of 640 acres. Each quarter-section contains 160 acres ($1/4 \times 640 = 160$). Knowing that 160 acres = one quarter-section and that a quarter-section of land is a tract $1/2$ mile (2,640 feet) on a side, determine:
   a) the number of square feet in a quarter-section (160 acres).
   b) the number of square feet in 1 acre.

9. Approximately how many acres are covered by Lake Bailey? (*Suggestion:* First determine the area of the lake in square feet by multiplying the length of the long axis by the length of the short axis. Do the same for the island. Subtract the area of the island from the area of the lake and convert the difference to acres.)

10. What is the direction of flow of the intermittent stream located in sections 10 and 11?

11. What is the elevation of the contour line that surrounds the small pond between Silver River and Highway 26 just south of the shore of Lake Superior? (Lake Superior is the large body of water across the northern part of the map.)

Figure 2.11 DELAWARE MAP
Part of the U.S.G.S. Delaware Quadrangle, Michigan, 1948. Scale, 1:24,000; contour interval, 20 feet.

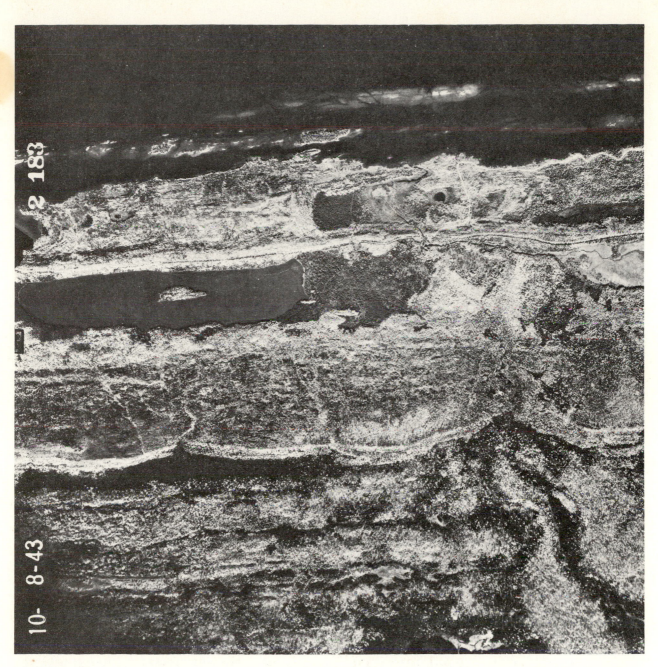

Figure 2.12 Aerial photograph of part of the Delaware Map on the opposite page. The photo is shown here for comparative purposes and will be used in a later exercise. (Courtesy of the U.S.G.S. Photo was taken on October 10, 1943.)

Aerial Photographs and Other Imagery from Remote Sensing

An aerial photograph is a picture taken from an airplane flying at altitudes as high as 60,000 feet. Practically all of the United States has been photographed aerially by federal, state, or other agencies.

Photographs taken with the axis of the camera pointed vertically down are *vertical photos* and are most useful to the geologist. Air photos are also used for the making of topographic maps, for highway location, military operations, mapping of soils, city planning, and for many other operations.

Other "pictures," made with special sensors mounted in earth-orbiting satellites, are also useful for displaying features of the earth's surface. The false color images are perhaps the most spectacular from an aesthetic point of view, but they are quite useful in geologic interpretative work because of the broad regional relationships they show. The false color images and other kinds of earth satellite imagery were made with imaging devices aboard the two Earth Resource Technology Satellites, Landsat 1 (launched July 23, 1972) and Landsat 2 (launched January 22, 1975). These two satellites circle the earth in a nearly polar orbit at an altitude of 570 miles (915 km). Each view of the earth's surface from the Landsat spacecraft covers a strip about 115 miles (185 km) wide. Landsat transmits the necessary data to receiving stations on earth. On the North American continent, there are three Landsat receiving stations in the United States and two in Canada. These data are processed into images of various kinds and made available to users through agencies of the federal government.

Space imagery results in "maps" that are very small in scale; that is, an inch on these image maps represents many miles on the ground. Thus, the Landsat maps do not have the resolution (degree of detail) that aerial photographs have. For this reason, most of the remote sensing imagery used in this manual will consist of aerial photographs. Where appropriate, a false color image will be used.

Determining Scale on Aerial Photographs

The scale of an air photo depends on both the focal length of the camera and the height of the airplane above ground surface. If these two factors are known, the R. F. of the photo can be determined. The photo scale can also be determined, however, by measuring known ground distances on the photo. In many parts of the coterminous United States, section lines are ideally suited for this purpose since they are normally one mile apart and are usually visible on the photo as roads, fences, or other man-made features. Conversion to an R. F. or graphic scale can be made in the same manner as was described under the section on topographic map scales.

Stereoscopic Use of Aerial Photographs

If two vertical photos are taken from a slightly different position and viewed through a *stereoscope,* the relief of the land becomes visible. A *stereopair* consists of two photos viewed in such a way that each of the eyes sees only one of the two photos. The brain combines the two images to form a three-dimensional view of the objects shown on the photograph.

Figure 2.13 is a stereopair for practice in using the simple lens stereoscope. Figure 2.14 shows two kinds of lens stereoscopes, and figure 2.15 shows one being used. The stereoscope is positioned over the stereopair so that the viewer's nose is directly over the line separating the two adjacent photos. If stereovision is not immediately achieved, the stereoscope should be rotated slightly around an imaginary vertical axis passing through the midpoint between the two lenses until the image viewed appears in relief.

On photographs in which steep slopes occur (e.g., the walls of deep canyons or the flanks of high mountains), the stereoscopic image of these features is exaggerated. That is, steeply inclined canyon walls may appear to be nearly vertical when, in fact, they are not. This distortion, however, does not present a problem under normal conditions of viewing by the beginning student.

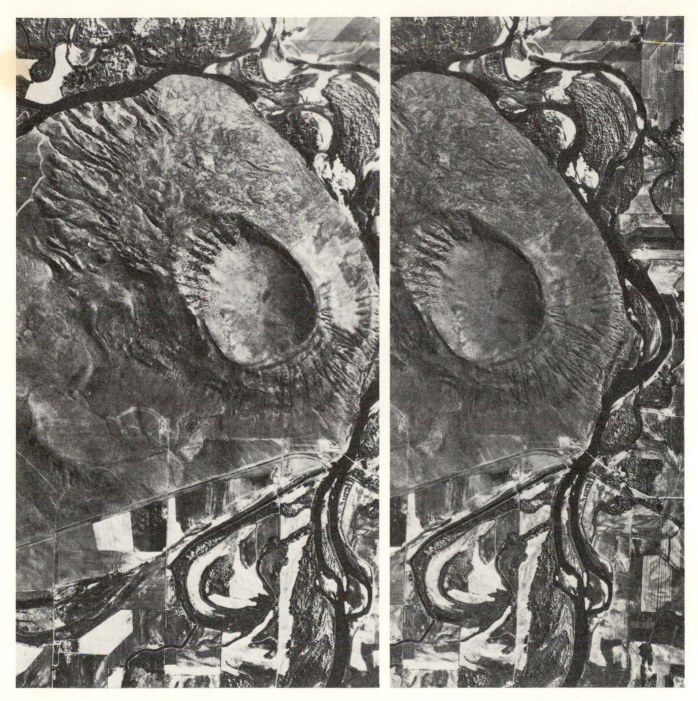

Figure 2.13   Stereopair of Menan Buttes, Idaho. The Snake River flows around the east side of the crater. (Courtesy of the U.S.G.S. Photos were taken on October 8, 1950.)

In both Parts 3 and 4 of this manual, several exercises will require interpretation of single aerial photos and stereopairs with reference to the terrain features and geologic phenomena shown on them. In preparation for those exercises, the paragraphs that follow will provide the student with some basic information on the interpretation of aerial photographs in terms of the recognition of common natural and man-made features.

Interpretation of Aerial Photographs

Air photos may be used to great advantage by geologists. Stereopairs are preferable, but single photos taken under good conditions of lighting and from proper altitudes reveal exactly what the human eye sees except for the third dimension, and even this can, with practice, be approximated from single photos.

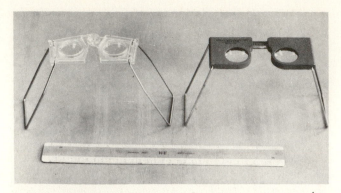

Figure 2.14 Two examples of a lens stereoscope. The one on the left is plastic and the one on the right is metal. Both can be adjusted with respect to the distance between the lenses, and both are collapsible. Scale is 12 inches long.

Figure 2.15 Student using a lens stereoscope to view a stereopair. In actual practice, the nose is positioned in the slot between the two lenses in order to bring the eyes closer to the lenses.

The interpretation of air photos is an art acquired only after considerable experience in working with photos from many different areas. However, beginning students can comprehend air photos amazingly well if they have a few simple instructions to follow and some basic principles to guide them.

The greatest difficulty confronting an individual looking at an air photo for the first time is recognition of familiar features that are seen every day on the ground but become mysterious objects when seen from the air. It is necessary, therefore, to acquire the ability to recognize certain common features before one can expect to use an air photo as a geologic tool. Some of these common features are described in the paragraphs that follow and are illustrated in figure 2.16.

Vegetation

Vegetational cover accounts for a great many differences in pattern and shades of gray tone on air photos. Heavily forested areas are usually medium to dark gray, whereas grasslands show up in the lighter tones of gray. Planted field crops are extremely varied in tone, depending not only on the kind of crop but also on the stage of growth. Cultivated fields are usually rectangular in shape and appear either in dark gray or light gray, depending upon whether the fields have just been plowed or whether the crop is already in and growing.

Soil and Rock

Soil texture controls the soil moisture, which in turn controls the appearance of the soil on air photos. Wet clayey soils have a much darker shade of gray than do the dry sandy soils, which usually show up as light gray to almost white in arid regions. Different tones of gray that show up in the same field are due to different degrees of wetness of the soil, a condition usually related to topographic low (wet) and high (dry) areas.

Where bedrock crops out at the surface, air photos reveal differences due to lithology, texture, mineral composition, and structure of the rocks. The beginner cannot always discern between a sandstone and a limestone, for example, but should have less trouble recognizing the differences among the major rock groups (i.e., igneous, sedimentary, and metamorphic). Vegetational patterns on air photos commonly reflect the underlying bedrock, and this is helpful in tracing out a single rock unit of the photo. On the other hand, if a thick mantle of unconsolidated material, either residual or transported, covers the area, then all bedrock features may be partially obscured or entirely obliterated.

Air photos used in connection with field investigations on the ground are among the most valuable tools of the professional geologist. When direct field examination is not feasible, air photos provide even better information than one could acquire by flying over the area in person, especially if the photographs are available for stereoscopic study. For the student of elementary physical geology, aerial photographs are useful in that they show geologic features of the earth's surface that otherwise would be difficult to describe or impossible to illustrate.

a. Sedimentary rocks uplifted to form dome (egg-shaped feature) in a semi-arid climatic zone. Stream valley with deciduous trees in valley bottom (black). Wyoming. Scale, 1:80,000.

b. Deeply incised river in flat lying sedimentary rocks. Badlands topography on either side of the river. Colorado. Scale, 1:80,000.

c. Small town in midwest. Various cultural features shown. Deciduous vegetation in town and along river. Missouri. Scale, 1:20,000.

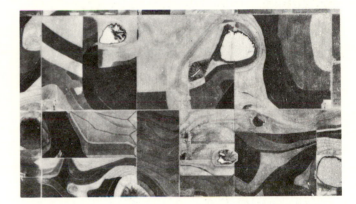

d. Agricultural field patterns (light and dark varigated area) with undrained depressions forming ponds (white due to reflection of sun off water surface). Contour plowing evident. Texas. Scale, 1:63,360.

e. Crystalline rocks with sparse coniferous vegetation. River in deeply cut valley. Lake (black) at margin of photo. Joint pattern in rocks evident. Wyoming. Scale, 1:60,300.

f. Glacial moraine from continental glacier. Kame and kettle topography with numerous kettle lakes (black). Rectangular field patterns in varigated colors. North Dakota. Scale, 1:60,000.

Figure 2.16 Some examples of features visible on aerial photographs.

Landsat False Color Images

The normal human eye can perceive a continuous spectrum from blue and green to yellow, orange, and red. Neither ultraviolet nor infrared radiation is visible to humans. The imaging technology used on the Landsat remote sensors records infrared radiation that, when combined with other wavelengths in the processing of the imagery data, results in a colored image in which the true colors are replaced by other colors. Hence the term, *false color image*.

In false color images, green vegetation shows up as various shades of red, deserts and other nonvegetated tracts are light grey to bluish grey, cities and large metropolitan areas are dark grey, and clear water areas are commonly black. Water bodies containing silt and other suspended sediments appear in various shades of light blue. Because vegetational patterns reflect to some extent the underlying soil and rock types, the various shades and hues of red and other colors on a false color image define the different kinds of soil and rock types.

Landsat views do not cover a rectangular area due to the fact that the earth rotates as the satellite passes overhead. This phenomenon results in a rhombohedral "picture" such as the image shown in figure 2.17. North is generally toward the top margin of the image, but a true north-south line must be determined independently of the image margins. Where visible, agricultural fields are useful for this purpose because their boundaries are commonly oriented in north-south and east-west directions. If these are not present, other features such as a coastline or a major river can be compared with a map showing the same features to determine true north.

Figure 2.17 is a false color image of part of the Atlantic Coast of the United States showing Long Island, the Hudson River, and a part of New Jersey. The Atlantic Ocean appears black in color, and the light-colored patches in New Jersey are cultivated fields. The brownish area in the southern part of the image is the forested coastal plain of New Jersey, and the dark area in the extreme northwest corner is part of the northern Appalachian Mountains. New York City and Newark, New Jersey at the mouth of the Hudson River are dark blue grey. The light-colored narrow strips of coastal beaches off the southern shore of Long Island and off the coast of New Jersey reflects the fact that these features are sandy and generally lack heavy vegetation.

References

Avery, Thomas E. 1977. *Interpretation of aerial photographs.* 3d ed. Minneapolis: Burgess Publishing Co. 392 pp.

Lattman, L. H., and Ray, R. G. 1965. *Aerial photos in field geology.* New York: Holt, Rinehart and Winston. 221 pp.

Kroeck, Dick. 1976. *Everyone's space handbook.* Pilot Rock, Inc. P. O. Box 470, Arcata, CA. 95521. 175 pp.

Williams, R. S., and Carter, W. D. eds. 1976. *ERTS-1 a new window on our planet.* U.S.G.S. Professional Paper 929, Washington, D.C. 362 pp.

Figure 2.17   False color image of the Long Island–New Jersey area made from Landsat 2 on October 21, 1975. The long narrow strip of sand beach off the Southern shore of Long Island is about 80 miles long. (NASA ERTS E–2272–14543, U.S.G.S. EROS Data Center, Sioux Falls, South Dakota 57198.)

Exercise 11. Introduction to Aerial Photograph Interpretation

1. Figure 2.18 is an aerial photograph of standard size. Identify the following features on the photograph. Where a topographic map symbol is available for the feature (refer to fig. 2.2), draw the symbol directly on the photograph of figure 2.18 at the place where the feature occurs. If no symbol exists, draw the outline of the feature on the map and label it accordingly. Use a red pencil in all cases except for water features, which should be shown in blue.
   a) Major road
   b) Railroad track
   c) River
   d) Airfield
   e) Football field and track
   f) Small stream valley
   g) Bridge over river
   h) Overpass
   i) Golf course

2. The southeast corner of the golf course is also the SE corner of Section 12, T. 92 N., R. 52 W. The "T" intersection of the road at the south edge of the golf course about 4 inches to the west (left) on the photograph is at the SW corner of the SE¼ of the SE¼ of Section 11, T. 92 N., R. 52 W. Determine the R. F. and verbal scale of the photograph.

3. Figure 2.19 is a stereopair. Examine it with a stereoscope and answer the following questions while completing the instructions.
   a) Does the major stream channel contain any signs of the presence of water?
   b) Trace the drainage lines in blue pencil on the right-hand photograph using the proper symbol from figure 2.2.
   c) What is the dominant vegetation of the area?
   d) Are any man-made features visible on the stereopair?
   e) The area covered by the stereopair shows two relatively flat upland surfaces away from the stream channel, each of which lies at a general elevation that differs from the other. Draw the boundaries of these areas on the right-hand photograph while viewing the stereopair with a stereoscope. Use a red pencil. Mark the lower-lying area with the number *1*, the higher area with the number *2*. Extend these boundaries on the rest of the photograph.

4. Figure 2.12 covers part of the Delaware Map of figure 2.11. With red pencil, label the following features with the lowercase letter that appears before each name:
   a) Lake Superior
   b) Lake Bailey
   c) Agate Harbor
   d) Small lake in section 35
   e) Swamp area south of Mt. Lookout
   f) Mt. Lookout
   g) Trace the roads in sections 34, 35, and 36 that are shown by a dashed red line on the Delaware Map.

5. Trace the drainage system from the Delaware Map onto the aerial photo of figure 2.12. Use a blue pencil and label each creek or river that has a name.

6. Determine the scale of the aerial photo in figure 2.12.

7. Figure 2.6 is an aerial photograph of an area in Central Iowa. Section 6 of T. 90 N., R. 37 W. is labeled on the photograph.
   a) Using the information provided, determine the scale of the photograph.
   b) Draw the township and range lines on the photograph using red pencil. Draw in the section lines with black pencil. Number all sections in black pencil. Along the photograph margins record the township and range designations in red pencil using the conventional system shown in figure 2.5.
   c) Note the relationship between the section lines and the road pattern. Compare this relationship with that on figure 2.18. Suggest possible reasons for the differences.
   d) A railroad track crosses the area from a southeasterly to a northwesterly direction. Use the appropriate map symbol to show this feature on the photograph. (Draw directly on the photograph with black pencil.)
   e) The rectangular areas covering most of the area are croplands. They appear in various shades of grey to nearly black on the photo. Explain the differences.

8. The scale of a vertical photograph depends on the focal length of the camera used and the height of the airplane taking the photo above the ground. The R. F. is the focal length of the camera divided by the height of the plane above the ground. Determine the scale of a photograph taken by a camera with a 9-inch focal length from an airplane flown at a height of 35,000 feet above the ground. (Be careful about units.)

North

Figúre 2.18   Aerial photograph of area in South Dakota. (Photo No. VE–1JJ; exposed on June 19, 1968.)

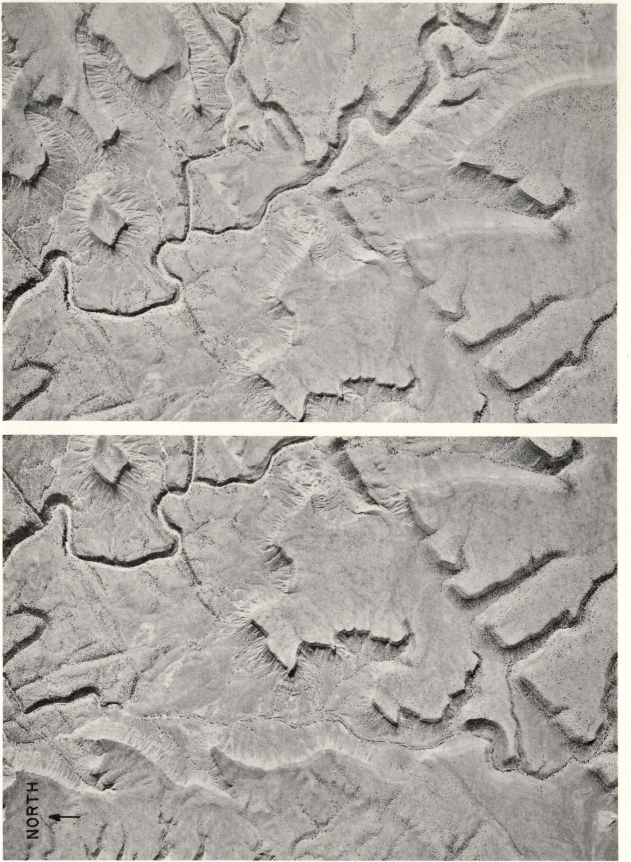

NORTH

Figure 2.19   Aerial photograph stereopair, Utah, 1956. Scale, 1:20,000. (Photograph numbers GS-RR-17-43 and GS-RR-17-44.)

Part 3

Geologic Interpretation of Topographic Maps, Aerial Photographs, and Earth Satellite Images

Introduction

In Part 2 of this manual we learned that the configuration of a landform is expressed on a topographic map by contour lines, and is revealed on aerial photographs when two overlapping photos are viewed stereoscopically. Topographic maps and stereopairs can thus be used for the study and analysis of terrain features in terms of the geologic processes that produced them. Images from earth-orbiting satellites are additional tools useful in the study of landforms.

Every geologic process leaves some imprint on the part of the earth's surface over which it has been operative. These processes include the work of the major geologic agents such as wind, groundwater, running water, glaciers, waves, and vulcanism. Each of these agents leaves its mark on the landscape in the form of one or more characteristic landforms.

The association of geologic agents with the origin of various landforms is a subdivision of the general discipline of geology called *geomorphology*. Geomorphologists have systematized the relationship of geologic processes to topographic forms into a body of knowledge that can be used in deciphering the origin of topographic features shown on a topographic map or seen in a stereopair. The body of knowledge that deals with the origin of landforms is presented in all basic textbooks that deal with physical geology.

It is assumed that the users of this manual will have become acquainted with the different geologic processes and their related landforms through reading appropriate chapters in a textbook on physical geology, and by listening to lectures in which the origin of landforms is presented. This is prerequisite to the understanding and successful completion of the exercises that are presented in this part of the manual.

General Instructions

The purpose of Exercises 12 through 19 is to acquaint the student with a variety of landforms and geologic principles associated with the geologic agents of wind, groundwater, running water, glaciers, waves, and vulcanism. Topographic maps and profiles, aerial photographs, some of which are in the form of stereopairs, and other pertinent maps, satellite images, diagrams, and data are provided in Exercises 12 through 19 as the basic tools for learning the association between landform and geologic agent, or to establish a specific geologic principle.

The title of each exercise identifies the geologic agent that will be under consideration for that particular exercise. It is assumed that you are thoroughly conversant with and have a good understanding of the material presented in Part 2 of this manual on topographic maps, satellite images, and aerial photographs, including map and photo scales, contour lines, map symbols, and topographic profiles. Moreover, it is necessary that you know the terminology associated with each of the geologic agents, because these terms are not always defined in the exercises in which they are used. *It is good practice to bring your textbook to the laboratory as a reference for unfamiliar or forgotten terms that crop up in the exercises.*

All maps and photographs needed for completion of the exercises are included in the manual. Before proceeding with the questions or problems based on maps or photographs, note the scale and contour interval on the topographic maps, and the scale of the stereopairs. Some exercises require you to draw on the maps or photos with ordinary lead pencil or colored pencils. In these cases, make your initial lines very light so that if erasure is necessary, it can be accomplished with ease. Also, use sharp pencils to insure accuracy, especially where the drawing of a topographic profile is a requirement of the exercise. Some students who suffer from eyestrain after prolonged study of maps may find an inexpensive magnifying glass helpful for completing the map exercises. The questions for each exercise should be answered in the order given because they are arranged in a more or less logical sequence.

Exercise 12. Geologic Work of Running Water

12A. PORTAGE MAP, Montana

A segment of the Missouri River flows across the map area of figure 3.2. Figure 3.1*A* shows the longitudinal profile of a riverbed. *Base level* is the lowest level to which a river can erode its bed. In figure 3.1*B*, another river profile has been interrupted by a dam. The water level in the reservoir above the dam forms a new base level for the segment of the river upstream from the dam. Below the dam, the river erodes a new profile that is controlled by the same base level that controlled the entire riverbed profile before the dam was constructed. Notice the sites of deposition and erosion above and below the dam in figure 3.1*B*. Refer now to figure 3.2.

1. What is the direction of flow of the Missouri River?

2. Three dams are located along the course of the Missouri River. On the basis of man-made installations associated with each dam, what purposes do the dams serve?

3. How many new base levels have been created by these dams?

4. In the stretch of the Missouri River between the Rainbow and Ryan dams, is the water depth greater near the Rainbow Dam or the Ryan Dam?

5. In the same stretch of river, state whether the riverbed is subject to erosion or deposition at the following sites:

    a) Immediately downstream from the Rainbow Dam.

    b) Immediately upstream from the Ryan Dam.

6. If the Ryan Dam were destroyed by an earthquake, describe the environmental effects on

    a) The channel of the Missouri River between Ryan Dam and Rainbow Dam.

    b) Morony Dam.

7. Beginning about 1 mile downstream from Rainbow Dam, a series of contour lines cross the Missouri River. In a distance of 1.7 river miles, the river drops 60 feet. On the basis of this information, what is the gradient of the river along this stretch? State your answer in feet per mile.

8. The Missouri River flows a total distance of 2,500 miles from its headwaters at 14,000 feet above sea level to its confluence with the Mississippi River at 410 feet above sea level. What is the average gradient of the Missouri River from source to mouth? Why does the average gradient differ from the gradient determined in question 7?

Figure 3.1   (*A*) Longitudinal profile of a river showing a gradual decrease in gradient downstream. Base level is the lowest to which a stream can erode its bed. (*B*) A longitudinal river profile that has been interrupted by a dam. The old, pre-dam base level has been replaced by a new one. (From Carla W. Montgomery, *Physical Geology.* Copyright © 1987 Wm. C. Brown Publishers, Dubuque, Iowa. All Rights Reserved. Reprinted by permission.)

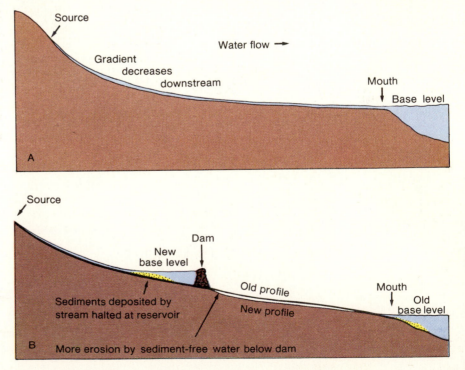

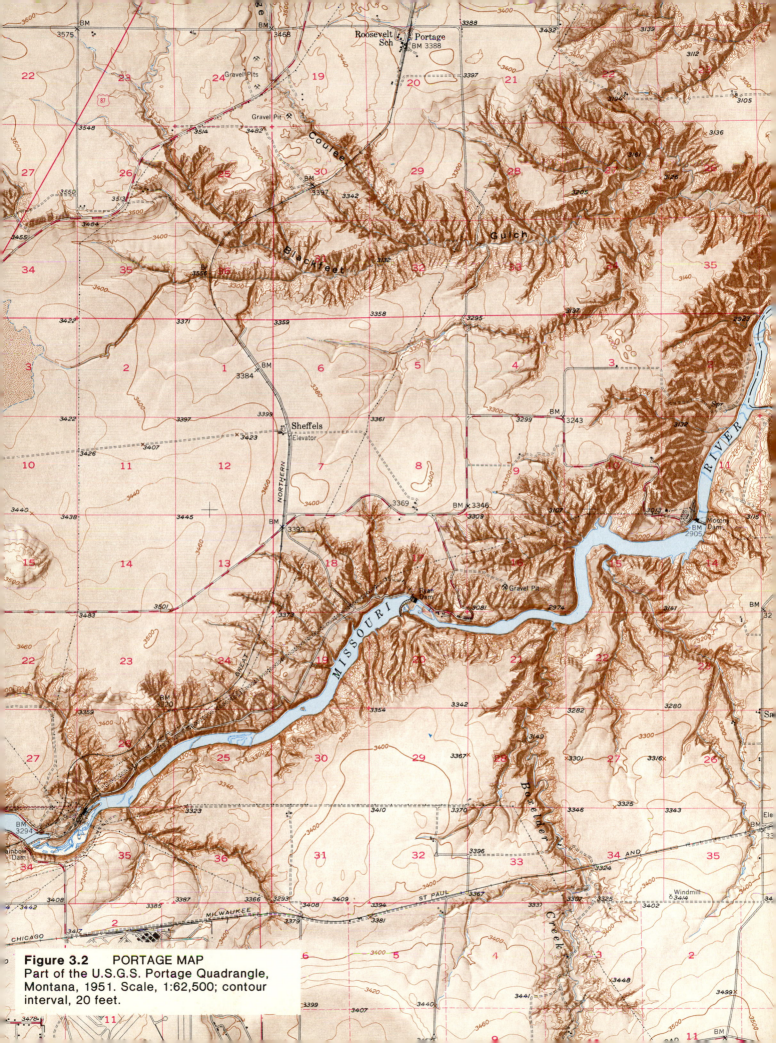

Figure 3.2   PORTAGE MAP
Part of the U.S.G.S. Portage Quadrangle,
Montana, 1951. Scale, 1:62,500; contour
interval, 20 feet.

12B. PROMONTORY BUTTE MAP, Arizona

This map (fig. 3.3) shows two stream systems, one flowing northward, the other southward. The Mogollon (pronounced Mo-gee-yone) Rim is a pronounced topographic feature that roughly separates the two. Inspection of the topography produced by the two systems reveals that the terrain south of the Rim is more rugged than the terrain north of the Rim. The purpose of this exercise is to analyze the difference between the two systems and evaluate the future of the Mogollon Rim as time passes.

The dominant geologic process both north and south of the Rim is stream erosion. The two opposing drainage systems are separated by a *divide*, a line connecting the high points between them. Figure 3.4 shows how a divide is shifted as headward erosion of two opposing drainage systems progresses. If the gradients of the streams on both sides of the divide are comparable, the divide will remain more or less fixed and simply be lowered as erosion continues. If, however, the gradients of the streams in the two systems differ appreciably, the divide should move progressively toward the system with the smaller gradients. We will now evaluate the change with time that can be expected in the position of the Mogollon Rim when this simple principle is applied.

1. Trace in red pencil the divide between the headward regions of the two stream systems. Use a soft black pencil first, then trace it in red when you are satisfied with its position.

2. What man-made feature is *roughly* coincident with the divide?

3. Does the Mogollon Rim lie on the divide or north or south of it?

4. Table 3.1 lists several streams north *(A)* and south *(B)* of the Mogollon Rim. For each stream, the elevations of two contour lines that cross the stream a few miles apart are recorded, the difference in elevation of the two lines is given, and the map distance between them is shown.
   a) For each stream segment listed in table 3.1, calculate its gradient in feet per mile.
   b) Determine the average gradient for the streams north and south of the Rim.

5. If the erosive potential of a stream is proportional to its gradient, which system, *A* or *B*, should have the greatest erosive power?

6. Will the divide between the two systems move generally toward the north, toward the south, or remain more or less fixed with the passage of time? Explain your answer.

7. Sketch a series of maps that show the progressive changes that will occur around Promontory Butte, especially at its juncture with the Mogollon Mesa.

Figure 3.3   PROMONTORY BUTTE MAP
Part of the U.S.G.S. Promontory Butte
Quadrangle, Arizona, 1951. Scale, 1:62,500;
contour interval, 50 feet.

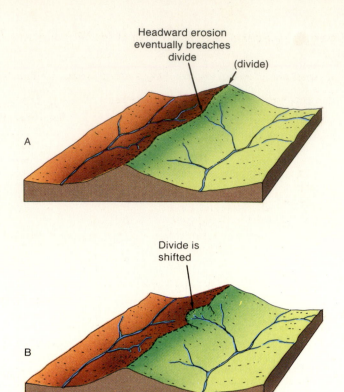

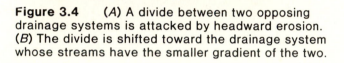

Figure 3.4 (*A*) A divide between two opposing drainage systems is attacked by headward erosion. (*B*) The divide is shifted toward the drainage system whose streams have the smaller gradient of the two.

Table 3.1. Matrix for Determining Stream Gradients North of the Mogollon Rim *(A)* and South of the Mogollon Rim *(B)*.

A. Determination of Stream Gradients North of Mogollon Rim, Promontory Butte Map, Arizona

Name of Stream	Difference in Elevation of End Points on the Stream Segment. High Point − Low Point = Diff. in Elev.			Length of Stream Segment in Miles	Gradient ft/mi
West Leonard Canyon	7,500'	7,250'	250'	2.6	
Middle Leonard Canyon	7,750'	7,350'	400'	2.4	
East Leonard Canyon	7,500'	7,250'	250'	1.9	
Turkey Canyon	7,700'	7,450'	250'	1.9	
Beaver Canyon	7,650'	7,350'	300'	3.7	
Bear Canyon	7,750'	7,550'	200'	1.6	
				Average Gradient	

B. Determination of Stream Gradients South of Mogollon Rim, Promontory Butte Map, Arizona

Name of Stream	Difference in Elevation of End Points on the Stream Segment. High Point − Low Point = Diff. in Elev.			Length of Stream Segment in Miles	Gradient ft/mi
Big Canyon Creek	7,000'	5,500'	1,500'	3.7	
Dick Williams Creek	7,000'	5,850'	1,150'	1.8	
Horton Creek	6,500'	5,500'	1,000'	3.0	
Doubtful Canyon	6,500'	5,250'	1,250'	2.4	
Spring Creek	6,750'	5,500'	1,250'	2.2	
See Canyon	6,700'	6,000'	700'	2.2	
				Average Gradient	

12C. ANTELOPE PEAK MAP, Arizona
ENNIS MAP, Montana

These two maps show two common landforms produced by running water. The Antelope Peak Map (fig. 3.5) shows a well-developed *pediment* and the Ennis Map (fig. 3.6) contains an excellent *alluvial fan*. Before proceeding, review the geologic origin of these two landforms by reading from your textbook or another reference.

1. On the grid of figure 3.7, draw a topographic profile of each of these features with the beginning and end points of the profile as described below. Begin each profile by plotting the *highest* elevation on the vertical axis at mile zero.

    a) Line of profile for the Antelope Peak pediment: the starting point is at the SW section corner of Section 35, T. 6 S., R. 2 E., at an elevation of 1,694 feet. The profile extends as a straight line for a little more than 6 miles in a northeasterly direction, passing through the SW section corner of Section 6, T. 6 S., R. 3 E., at an elevation of 1,320 feet, and ends at the 1,300-foot contour line about one-half mile northeast of the previously described point. Draw the line of profile on the Antelope Peak Map with a sharp pencil.

    b) Line of profile for the Cedar Creek Alluvial Fan, Ennis, Montana: the starting point is at the northeast section corner of Section 21 in the southeast part of the Ennis Map area (Section 21 is the one in which the Lawton Ranch is located). The elevation at the starting

point (i.e., mile-zero) is 6,080 feet above sea level. From this point, the line of profile extends as a straight line in a northwesterly direction for a distance of about 4 miles, passing through the northwest section corner of Section 7 (elevation 5,245), and ending at the 5,200-foot contour line. (In drawing this profile, it will be more convenient to turn the map upside down so that the starting point of the profile will be at your upper left.) Draw the line of profile on the Ennis Map with a sharp pencil.

2. Both the pediment and the alluvial fan are produced by running water. Describe each profile in terms of its shape (i.e., concave, convex, or straight) and gradient in feet per mile.

3. Inasmuch as both the pediment and the alluvial fan are produced by the action of running water, suggest a reason or reasons why the two profiles are different.

4. The uniform regularity of contour spacing on the pediment of the Antelope Peak Map is interrupted by several isolated hills that rise 100 to 200 feet above the level of the pediment between Mesquite Road and Highway 84. What is the origin of these features?

5. Ennis Lake was formed behind a dam on the Madison River (not shown on the map). What effect has the dam had on the segment of the Madison River shown on the Ennis Map?

Figure 3.5   ANTELOPE PEAK MAP
Part of the U.S.G.S. Antelope Peak
Quadrangle, Arizona, 1963. Scale, 1:62,500;
contour interval, 25 feet.

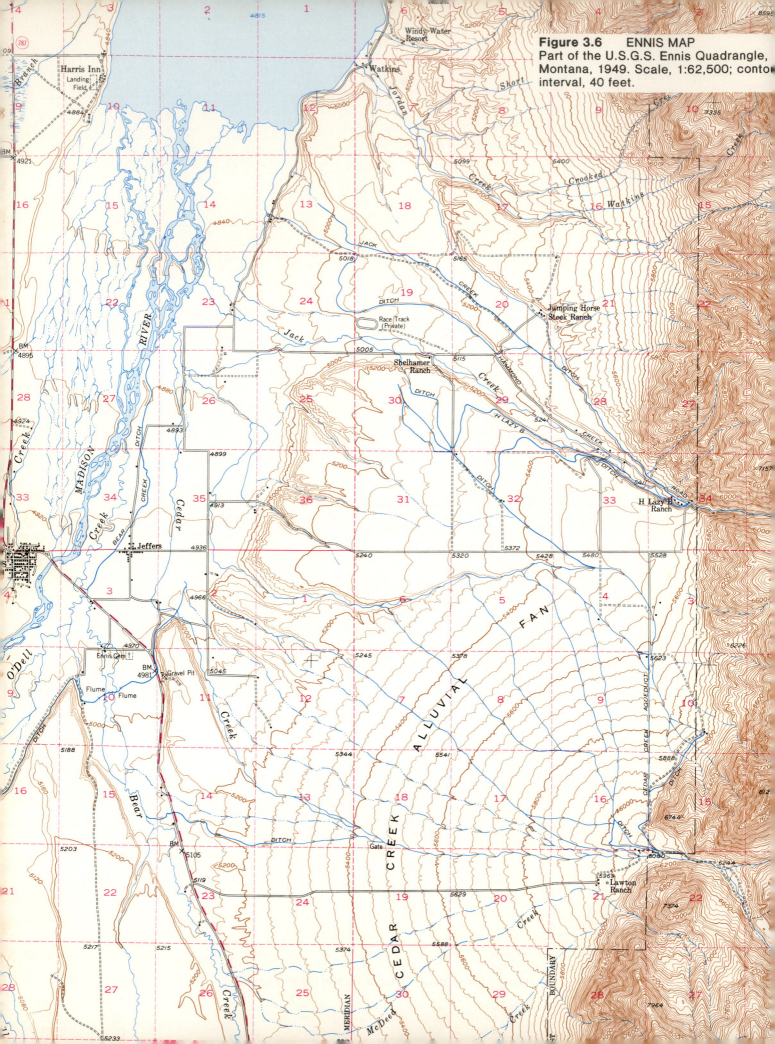

Figure 3.6   ENNIS MAP
Part of the U.S.G.S. Ennis Quadrangle,
Montana, 1949. Scale, 1:62,500; conto
interval, 40 feet.

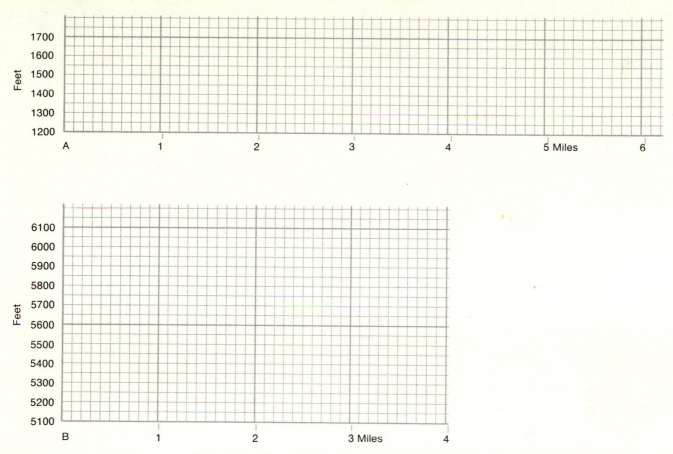

Figure 3.7 (*A*) Grid for drawing pediment profile based on the Antelope Peak Map. (*See* Exercise 12C-1a for location.) (*B*) Grid for drawing a profile of the Cedar Creek Alluvial Fan, Ennis Map. (*See* Exercise 12C-1b for location.)

Aluvial plain are usually in arid climates

12D. CHARLES CREEK, Canada

The aerial photograph (fig. 3.9) shows a meandering main stream and a single meandering tributary. Each has shifted its channel a number of times, both by channel migration and meander cutoffs. Channel migration occurs because of erosion by the river on the outside of a meander and deposition on the inside. Crescent-shaped sandbars on the inside bank of a meander are evidence of deposition. Meander cutoffs result when two meanders migrate toward each other causing the destruction of the narrow neck of land separating them as shown in figure 3.8. The isolation of a meander from the stream channel results in an oxbow lake. With the passage of time, the oxbow lakes become filled with sediment from floodwaters and aquatic vegetation.

1. Show by a series of diagrammatic map sketches how a U-shaped meander will be transformed into an oxbow lake. Indicate on your map the points of erosion by the letter *e*, and deposition by the letter *d*.

2. On the Charles Creek aerial photo (fig. 3.9), trace the present courses of the main stream and its tributary in blue pencil. Where the channel width is greater than the width of your pencil line, make your line follow the thread of maximum velocity as indicated by erosion on the outside banks of meanders and other bends in the river. Your pencil line, if drawn correctly, traces the *thalweg* of the river channel.

3. The point at which a tributary joins the main stream is called the *confluence*. Note the point of confluence at the present time. Indicate by a dashed red line the course of the tributary stream when its confluence with the main channel was different than at present.

4. Indicate by the letter *A* (with red pencil) the most recent meander cutoff on the main channel.

5. Indicate by the letter *B* (with red pencil) where the next meander cutoff on the main channel will occur.

6. Three former meander loops of the main channel are designated by the letters *x*, *y*, and *z* on figure 3.8. Which of these was most recently part of the river channel, and which has been an oxbow lake the longest? How determined?

7. A stream segment with many meanders has a gradient determined by the length of the stream channel and the difference in elevation of the end points of the segment. If one of the meanders in the segment becomes an oxbow lake, will the gradient remain the same, become higher, or lower? Use figure 3.8 as a reference in determining your answer.

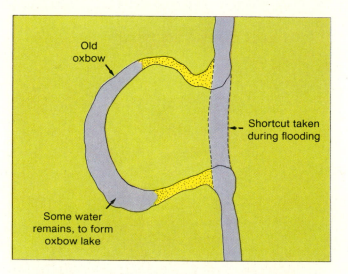

Figure 3.8   An oxbow lake forms when a river meander is separated from the main channel during flood stage. (From Carla W. Montgomery, *Physical Geology.* Copyright © 1987 Wm. C. Brown Publishers, Dubuque, Iowa. All Rights Reserved. Reprinted by permission.)

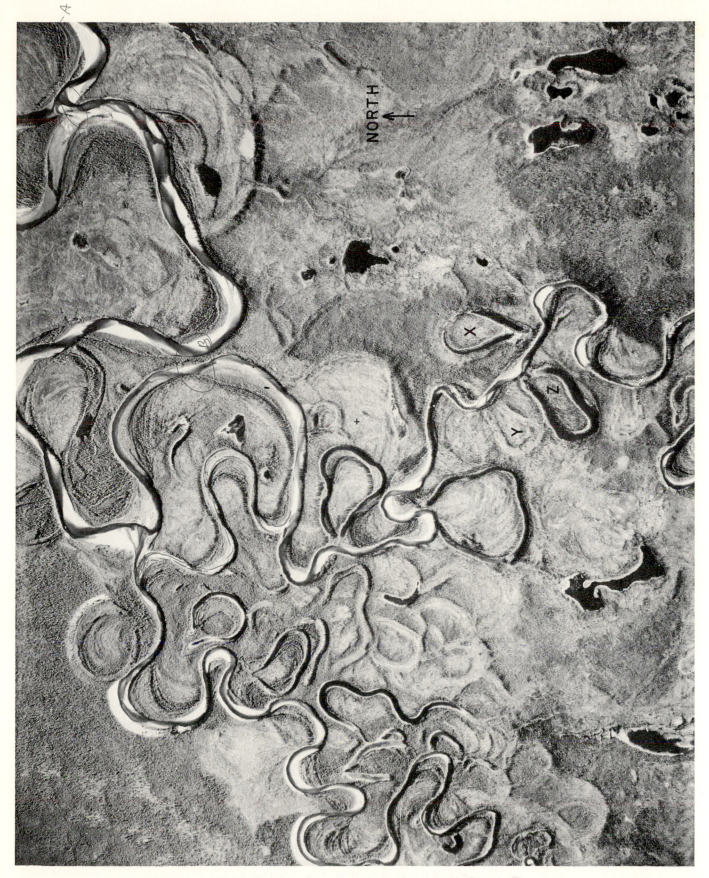

Figure 3.9 Aerial photograph of Charles Creek, south of Great Bear River, Northwest Territories, Canada. Scale, 1 inch = 1,200 feet. (By permission of the Royal Canadian Air Force.)

Interpretation of Maps, Photographs, and Satellite Images

12E. REFUGE MAP, Arkansas-Mississippi
GREENWOOD MAP, Arkansas-Mississippi

The Refuge Map (fig. 3.11) depicts a typical segment of the lower reaches of the Mississippi River. The course of the river in the map area is partly a natural one and partly an artificial one due to engineering works of the U.S. Army Corps of Engineers. These works include: artificial cutoffs, which are channels dug by the Corps in an attempt to straighten the meandering course of the river; and revetments, which are concrete slabs or other materials designed to reduce erosion on river banks and impede the migration of meanders. The entire map area lies on the floodplain of the Mississippi River. The origin of a floodplain by lateral erosion of a river is shown in figure 3.10.

The modern course of the river across the map area is from Miller Bend (near the north margin of the map) southward through the Tarpley cutoff (constructed in 1935) and the Leland cutoff (constructed in 1933).

Previous courses of the Mississippi River are marked on the map. One is defined by the *meander line* of 1823. It is shown on the map as a dotted line with the respective date printed alongside. Another previous course of the river is coincident with the boundary line between Arkansas (west of the Mississippi River) and Mississippi that lies generally to the east of the river. Notice, however, that certain parts of the state of Mississippi lie on the *west* side of the modern course of the river. This is the result of a court ruling that states where a stream or river forms the boundary between states, and the channel of the stream is changed by the ". . . natural and gradual process known as erosion and accretion, the boundary follows the varying course of the stream . . . while if the stream from any cause, natural or artificial, suddenly leaves its old bed and forms a new one by the process known as avulsion, the resulting change of channel works no change of boundary, which remains in the middle of the old channel . . . although no water may be flowing it in."*

1. Draw the following lines on the Refuge Map:
   a) The course of the river at the time the Ark.-Miss. state boundary was established. (Trace the boundary as shown with red pencil. This line roughly corresponds to the *thalweg,* a line joining the deepest points of the river at that time.)
   b) The course of the river during the year 1823 (use blue pencil).

2. Study the two courses and deduce whether the Ark.-Miss. boundary was established before or after the 1823 course. (Meanders tend to migrate in a down-valley direction.)

3. Figure 3.12 shows a segment of the Mississippi River along the border between the states of Arkansas and Mississippi. The course of the modern river is not coincident with the boundary line between the two states, which was fixed by Congress before the modern course of the river was established.

   We will refer to the course of the Mississippi River shown on figure 3.12 as the "Modern Course," and the course of the river when the Arkansas-Mississippi boundary was set as the "Boundary Course." From the northern to southern boundaries of the Greenwood Map, the Modern Course is 65 miles and the Boundary Course is 136 miles.
   a) Express as a percent the amount of shortening of the Modern Course compared to the Boundary Course.
   b) What is the impact of this shortening on the *competence* and *capacity* of the river?

4. Draw the boundaries of the Refuge Map on the Greenwood Map. The Refuge Map was published in 1939 and the Greenwood Map in 1965. Compare the Refuge Map with its corresponding area on the Greenwood Map. Describe the changes in former or modern channels that have occurred in this 26-year period with particular reference to the following:
   a) Areas around Linwood Neck, Luna Bar, and Point Comfort.
   b) The course of the Tarpley Cut-off with respect to the dike on its east side.

*U.S.G.S. Bulletin 817 (1930). *Boundaries, areas, geographic centers and altitudes of the United States and the several states,* p. 3.

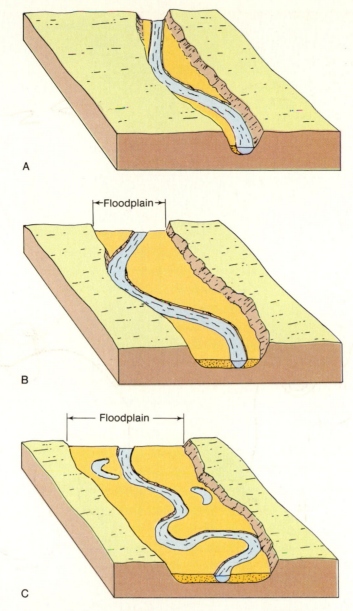

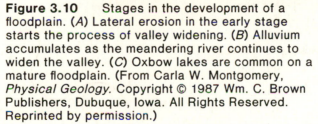

Figure 3.10 Stages in the development of a floodplain. (*A*) Lateral erosion in the early stage starts the process of valley widening. (*B*) Alluvium accumulates as the meandering river continues to widen the valley. (*C*) Oxbow lakes are common on a mature floodplain. (From Carla W. Montgomery, *Physical Geology.* Copyright © 1987 Wm. C. Brown Publishers, Dubuque, Iowa. All Rights Reserved. Reprinted by permission.)

Figure 3.11 REFUGE MAP
Part of the U.S.G.S. Refuge Quadrangle,
Arkansas-Mississippi, 1939. Scale, 1:62,500;
contour interval, 5 feet.

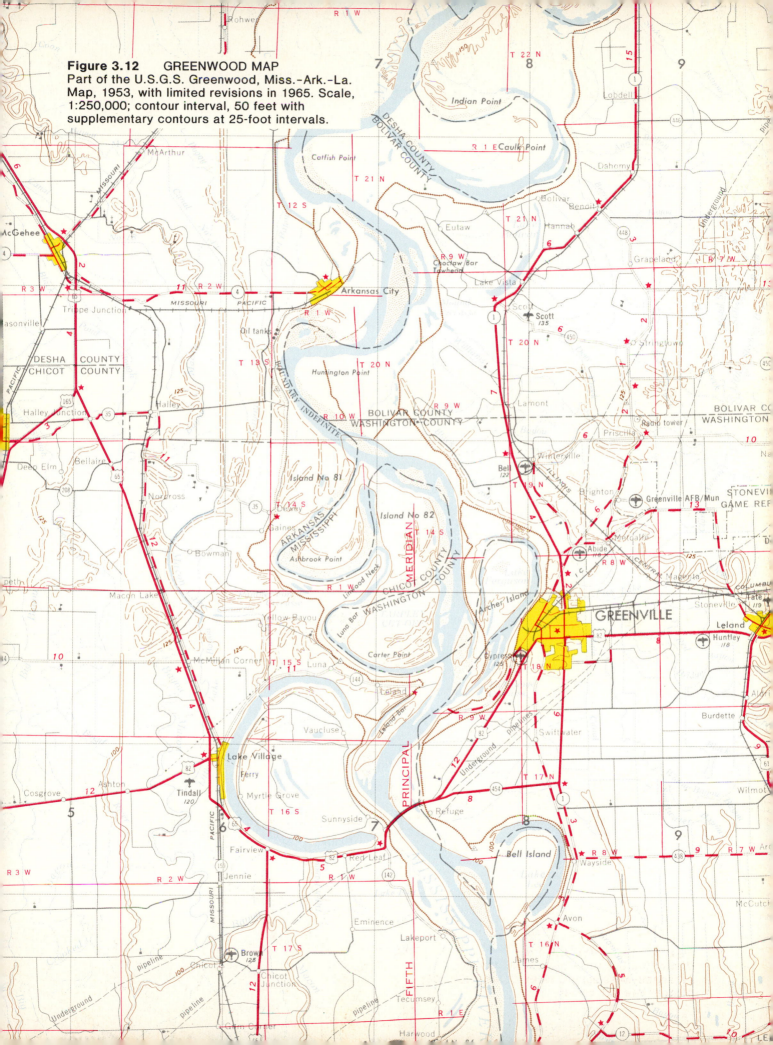

Figure 3.12   GREENWOOD MAP
Part of the U.S.G.S. Greenwood, Miss.-Ark.-La.
Map, 1953, with limited revisions in 1965. Scale,
1:250,000; contour interval, 50 feet with
supplementary contours at 25-foot intervals.

Groundwater

Groundwater Movement

Groundwater is the water that occurs beneath the surface of the earth in the *zone of saturation.* The top of the zone of saturation is the *water table.* Groundwater originates as rain that percolates downward through porous soil and rock until it reaches the water table. The water table lies at variable depths below the surface of the earth.

The water table is a planar feature and its configuration can therefore be defined by contour lines in the same way that the configuration of the earth's surface can be defined by contour lines on a topographic map. Contour lines on the water table can be drawn if enough points of known elevation on the water table are available for plotting on a base map. These points are assembled from water wells and other places of known elevation where the water table intersects the earth's surface, such as a stream or a lake.

Groundwater moves under the influence of gravity through porous rock or unindurated sands and gravels, but the movement is very slow compared to the velocity of flowing water in a river. Groundwater moves along flow lines. A *flow line* is a path followed by a water molecule from the time it enters the zone of saturation until it reaches a lake or stream where it becomes surface water.

Figure 3.13 is a map of a hypothetical area underlain by clean sand showing the relationship of the water table contours and flow lines to a permanent surface stream flowing south. For example, a water molecule at point *a* will follow the path of the flow line until it enters the stream at *a'*. The same relationship holds true for water molecules at *b, c, d,* and *e.* They follow the flow lines as they move down the slope of the water table at right angles to the water table contours until they enter the stream at points *b', c', d',* and *e',* respectively.

Flow lines can converge or diverge but they cannot cross each other. Moreover, as can be deduced from figure 3.13, groundwater from the west side of the stream cannot move across the stream to be comingled with groundwater on the east side of the stream, and vice versa, because the stream intercepts the flow of groundwater from both sides. From this simple analysis, it follows that any pollutant introduced into the zone of saturation will be carried by the flow of groundwater along flow lines until it eventually is discharged into a lake or stream.

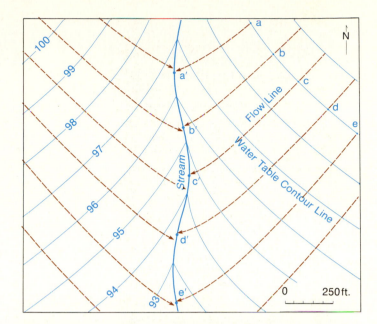

Figure 3.13   Map of a hypothetical area underlain by a well-sorted coarse sand showing a south-flowing permanent stream and contours on the water table. The ground surface is roughly the same shape as the water table but about 5 to 10 feet higher. The contour interval of the water table contours is 1 foot. Flow lines are shown in dashed colored lines. (See text for further explanation.)

Conversely, if a pollutant were introduced into the stream of figure 3.13 at point *a',* it would not enter the water table but would flow down the stream channel in the direction of *b'.* This relationship between groundwater flow and stream flow holds only for the case of an *effluent stream,* which is *a stream fed by groundwater;* that is, the groundwater is discharged into the stream channel. An *influent stream,* on the other hand, is one in which *the groundwater flows away from the stream channel.* In such cases, pollutants dumped directly into the stream will move into the zone of saturation. Influent streams also are commonly *intermittent streams,* streams that flow only during certain times of the year when rainfall is insufficient to supply surface runoff directly to them. With these basic principles in mind, we will now apply them to the following exercises.

Exercise 13. Groundwater

13A. ASHBY MAP, Nebraska

This area (fig. 3.14) lies in the Sand Hills of western Nebraska. The Sand Hills are formed from ancient sand dunes that are now more or less stabilized by surface vegetation consisting mainly of native grasses, an environment that makes this an excellent area for the grazing of cattle.

Sand dunes are composed of windblown sand that is very porous. Rain striking the surface in the Sand Hills very quickly percolates to the zone of saturation where it becomes groundwater. The water table in the area covered by the Ashby Map is quite shallow (i.e., close to the surface) in the interdune "valleys," which explains the many lakes and marshes that occur there.

The numbers imprinted on some of the lakes in this map area are elevations of their water surfaces. For example, the water surface elevation of Castle Lake is 3,767 feet above sea level. Assuming that these elevations are points on the water table, it is possible to draw a rough approximation of the water table contours. Some lakes have no elevations marked on the map. In these cases, the lake elevation can be estimated by noting the elevation of the contour nearest to the lake shore. Careful inspection of the map reveals that the lowest known lake elevation is North Twin Lake in the northeast corner of the map (3,737 feet above sea level), and that the highest lake elevation is Melvin Lake in the northwest corner of the map area (3,798 feet above sea level). From this relationship we can deduce that the water table must be sloping downward in an easterly direction. Therefore, the contour lines on the water table must extend at right angles to the direction of slope of the water table, or in a general north-south direction.

1. With a soft pencil (easily erasable) sketch in the water table contours for the entire map area. Use a C.I. of 10 feet. Start with the 3,800-foot contour line that begins just west of Melvin Lake and extends in a southerly direction until it passes just west of Hibbler Lake (about 2 miles northeast of the town of Ashby) with an elevation of 3,796 feet above sea level. Draw all water table contours on the map. For best results work first on the contours in the south-central part of the map area where more lakes are present, hence more control points are available. (Be sure that each water table contour is labeled.)

2. Locate the point in Section 20 (south half of the map) marked with a brown *x* and the numerals 3,862, which is the surface elevation at that point. On the basis of the water table contours that you have drawn on the map, estimate the depth of the water table at this point.

3. Suggest a reason why there are so few surface streams in this area.

4. The many flowing artesian wells in the map area indicate that two groundwater regimes are present. One is the water table regime defined by the water table contours that you have already drawn. The other is an artesian system, the existence of which is proved by the flowing wells. The *potentiometric surface* of an artesian aquifer is defined by the elevations to which water will rise in a number of wells that tap an artesian aquifer. If the elevation of the potentiometric surface is higher than the elevation of the ground surface, a flowing artesian well can be expected at that point.

Most of the flowing wells shown on the Ashby Map have black numbers printed next to them. This number indicates the elevation of the *ground surface* at the point where the flowing well is located.

On the basis of the foregoing, why is it not possible to draw contour lines of the potentiometric surface associated with the flowing wells in the map area?

Reference

Sniegocki, R. T. 1959. Geologic and groundwater reconnaissance of the Loup River Drainage basin Nebraska, *U.S.G.S. Water-Supply Paper 1493*. Washington, D.C.: U.S. Government Printing Office. 106 pp.

thalweg - deepest channel in a stream.
Competence - largest particle size carryable
Capacity - amt/unit time
gradient = Δ Elevations
              Length Stream Channel

Qz #5)
P. 72 4 a+b
         5
         6
P. 80 3-7
   82 3a+b

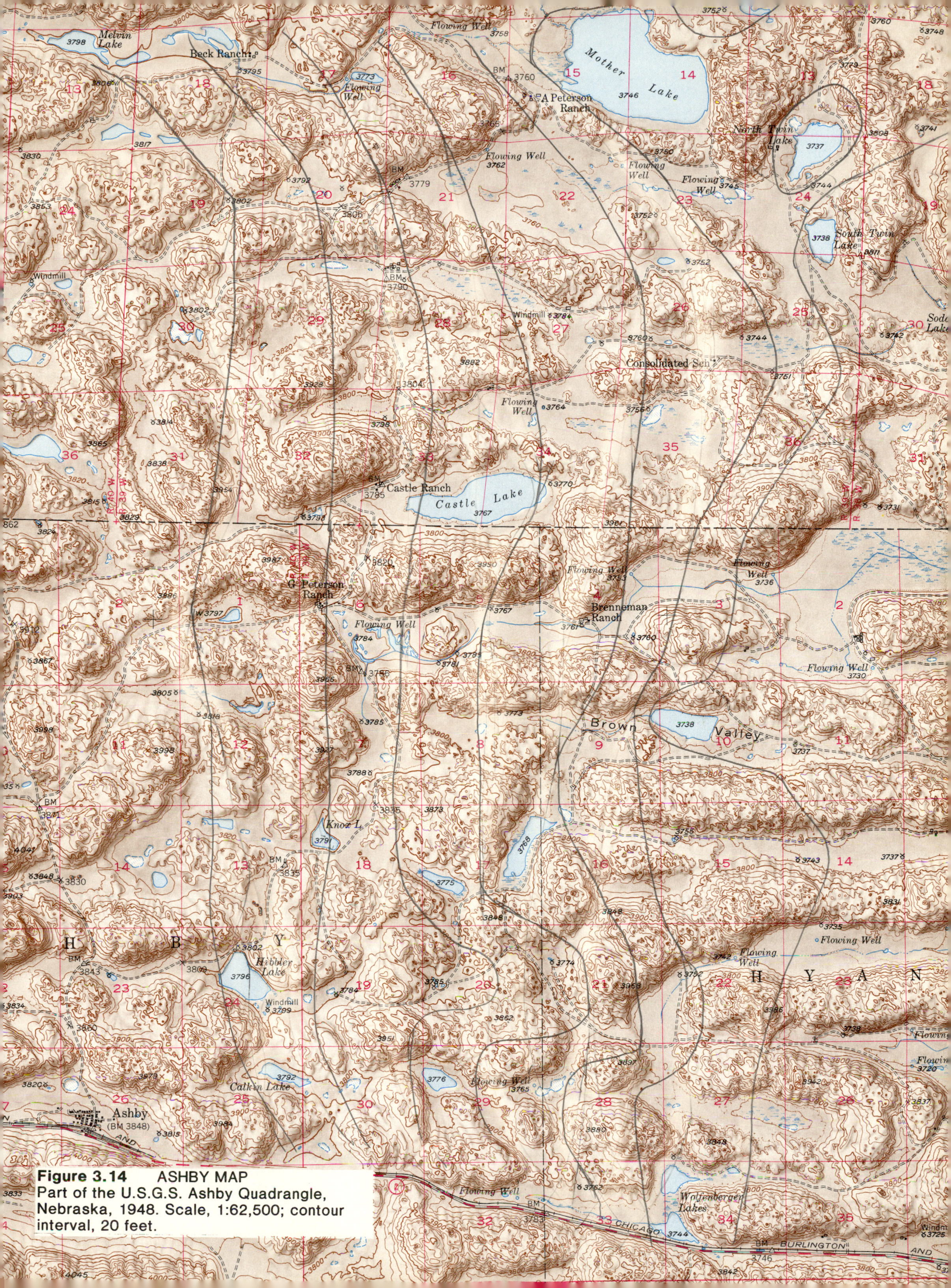

Figure 3.14   ASHBY MAP
Part of the U.S.G.S. Ashby Quadrangle,
Nebraska, 1948. Scale, 1:62,500; contour
interval, 20 feet.

13B. GROUNDWATER POLLUTION

Figure 3.15 is the map of a hypothetical area underlain by a well-sorted coarse sand crossed by a permanent stream, Clear Creek, that flows in a southeasterly direction. The ground surface is gently sloping to the southeast, and the shallow water table is defined by the water table contours. The location of a dump is shown on the map. The dump is privately owned and operated by a small company that hauls trash and garbage for residents of a nearby small town. The dump is an excavated pit, the bottom of which lies just above the water table.

The Jones property lies southeast of the dump, and its west property line borders on Clear Creek. Mr. Jones owns horses that he keeps in the barn and corral part of the time. The rest of the time the horses graze on the property and occasionally drink the water from Clear Creek.

The Smith estate lies west of Clear Creek and also has frontage on the creek. Both Jones and Smith derive their domestic water supplies from wells that penetrate the shallow water table. To assure themselves that the water was suitable for human consumption, Jones and Smith had well-water samples analyzed by the county health department at the time their wells were completed. Both Smith and Jones owned their respective properties for many years prior to the establishment of the dump, and have enjoyed a potable water until recently.

Recently, Jones's well water began to deteriorate in quality. This was verified when Jones had his water tested again at the county health department. Jones attributed this to pollution of the groundwater from leaching of domestic waste deposited in the dump. In talking with his neighbor, Smith, Jones suggested that the two of them should file suit against the owner of the dump and obtain a court injunction that would require cessation of all further dumping.

So Smith had water from his well tested again and found that it had not changed in quality since the tests conducted prior to the creation of the dump. Yet stream samples along the stretch that forms the boundary between the Jones and Smith properties were also analyzed by the county health department and were found to be contaminated by materials similar to those found in the recent water samples from the Jones well. This was sufficient evidence to convince Smith that he

ought to join in the suit with Jones against the dump operator. Smith reasoned that if the creek was contaminated with the same materials found in the Jones well, it would be only a matter of time until his well would also be polluted.

At that point, Jones and Smith hired an attorney to file the suit. He sought the advice of a geologist at a nearby university who had access to publications of the U.S.G.S. in the school's library. There he discovered a general geology report of the geology of the area. A section of the report on groundwater contained a map showing water table contours based on static water levels in other wells not shown in figure 3.15. The geologist transferred those contours to a map he was preparing for the lawyer. This map is figure 3.15.

1. On the map of figure 3.15, several dots are shown along the 800-foot water table contour line. Assume that these are the points of intersection of flow lines with the water table contours. Using the relationship of flow lines to water table contours as shown in figure 3.13, sketch a network of flow lines on figure 3.15. One flow line should pass through each of the dots on the 800-foot contour line. (Use a soft black pencil because you may have to erase several times before you are satisfied with your results.) Extend the flow lines across the entire map area so that the movement of groundwater can be ascertained.

2. On the basis of the flow line network that you have constructed on figure 3.15, answer the following questions:
   a) Is there reasonable evidence to conclude that seepage from the dump has contaminated the Jones well? Explain.
   b) Is there reasonable evidence that the stretch of Clear Creek adjoining the Jones and Smith properties has been contaminated by seepage from the dump? Explain.
   c) Is there reasonable evidence that the Smith well will be contaminated by seepage from the dump at some time in the future? Explain.

3. Is it possible that the animal waste in the corral on the Jones property is responsible for polluting the Jones well, any part of Clear Creek, or the Smith well, eventually? Explain.

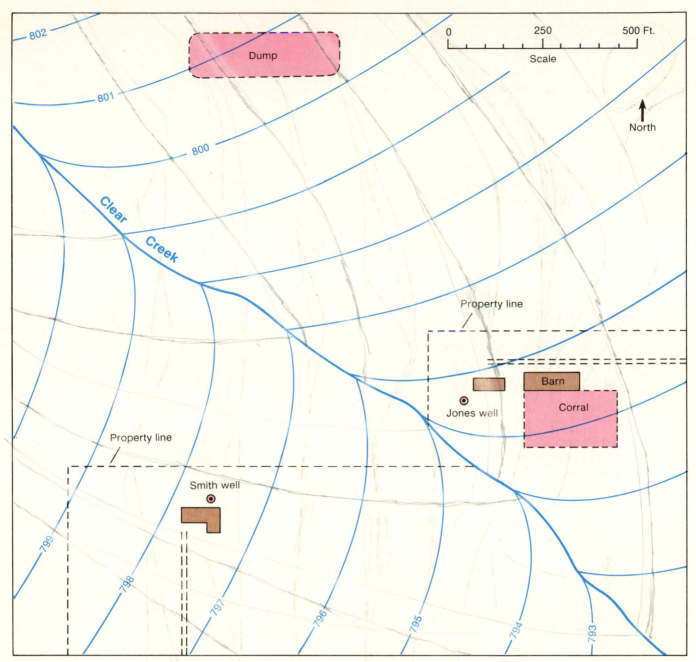

Figure 3.15   Map of a hypothetical area underlain by a well-sorted coarse sand about 50 feet thick. Clear Creek flows to the southeast. The water table contour interval is 1 foot. The water table lies about 8 to 10 feet below the ground surface, except near Clear Creek where the water table becomes shallower until it intersects the creek. The dump is an excavated pit, the bottom of which does not quite reach the water table.

Karst Topography

Groundwater slowly dissolves such rocks as limestone, dolomite, and rock salt. This process of *groundwater solution* forms caves and connecting passageways. Ultimately, the roofs of these underground cavities collapse leaving surface depressions called *sinks* or *sinkholes*.

A terrain marked by many sinks is called Karst topography, a name first applied to a limestone region along the Dalmation coast of Yugoslavia. In the United States, two areas where Karst topography is well developed lie in Florida and Kentucky. The Lake District in central Florida consists of thousands of water-filled sinks, and the famous Mammouth Cave of Kentucky is part of a system of interconnecting caverns and passages formed in limestone.

Exercise 14. Karst Topography

14A. INTERLACHEN MAP, Florida

Many of the sinks in the area covered by the Interlachen Map (fig. 3.16) are identified by concentric depression contours. Some of these sinks contain lakes.

1. A pair of sinks straddles the boundary between sections 5 and 8 about 1½ miles north of the east-west highway that crosses the central part of the map area. What is the approximate elevation of the bottoms of these two sinks?

2. In the same general area, an unnamed lake occupies the sink in the NE¼ of Sec. 7, and another occurs in the SW¼ of Sec. 4. Assuming that the water surface of these two lakes lies 5 feet below the elevation of the contour nearest the lake shores, what is the elevation of these two lakes? (Refer to them as Section 7 Lake and Section 4 Lake.)

3. Why are the two sinks in sections 5 and 8 dry—that is, they contain no lakes—when Section 7 Lake and Section 4 Lake both lie less than a mile away?

4. Generally speaking, the northern half of the area has greater relief than the southern half. The northern half has no streams while the southern half has Gun Creek, Cabbage Creek, and Little Cabbage Creek. Assuming that these two areas reflect different stages in the evolution of Karst topography, will the northern half become more like the southern half during the passage of time, or vice versa?

Figure 3.16 INTERLACHEN MAP
Part of the U.S.G.S. Interlachen Quadrangle, Florida, 1949. Scale, 1:62,500; contour interval, 10 feet.

14B. MAMMOTH CAVE MAP, Kentucky

The area covered by this map (fig. 3.18) is a good example of the effect of underlying strata on the topography. A cursory inspection of the map reveals that the southern one-third of the map area is pronouncedly different than the northern two-thirds from a topographic point of view. The reason for this is readily apparent from the geologic cross-section in figure 3.17. This north-south cross-section passes through the main road intersection in the town of Cedar Spring near the center of the map, and its horizontal scale is the same as the scale of the map. The pronounced change in topography on the map is marked by the Dripping Spring Escarpment lying just north of and roughly parallel to the Louisville and Nashville Road.

1. Draw the line of the geologic cross-section on the map in black pencil.

2. Generally speaking, the area north of the Dripping Spring Escarpment and *west* of the line of the cross-section is characterized by an integrated stream system. Use a blue pencil to trace the drainage lines occupied by permanent and intermittent streams. Use the correct map symbol for each. Take care not to extend your blue lines beyond the limits of the streams as they are shown on the map. What is the dominant bedrock in the area drained by the stream system?

3. Why does the stream flowing north into Double Sink end so abruptly there?

4. Examine the topography *east* of the line of cross-section and north of Dripping Spring Escarpment. A number of valleys such as Cedar Spring Valley, Woolsey Hollow, and Owens Valley resemble stream-cut valleys, but there are no streams flowing in them. Account for the fact that the topography of this area resembles a stream-dissected terrain even though no streams occupy the existing valleys.

5. In making a comparison of the terrains east and west of the line of the cross-section (fig. 3.18), which of the following statements is most likely the correct one?
   a) The area *west* of the line of cross-section will eventually resemble the area *east* of the cross-section as the streams cut deeper into the sandstone and encounter the underlying limestone.
   b) The area *east* of the line of cross-section will eventually resemble the area *west* of the cross-section as the overlying sandstone is further eroded.

6. Assume that the sandstone formation in figure 3.17 had an original thickness of 100 feet. On figure 3.17, draw the upper and lower contacts of the sandstone formation as it existed at some time in the geologic past before stream erosion began stripping it away.

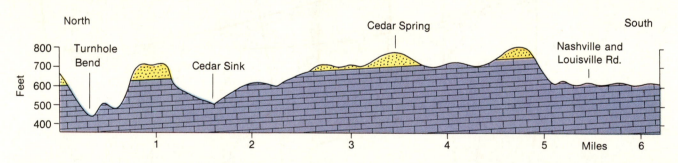

Figure 3.17 North-south geological cross-section from Turnhole Bend on the Green River to a point about three-fourths of a mile south of the Louisville and Nashville Road, Mammoth Cave Map. Geology is simplified from *The Geology of the Rhoda Quadrangle*, Ky., U.S.G.S. Map GQ-219 (1963). Vertical scale exaggerated about 10 times.

Figure 3.18   MAMMOTH CAVE MAP
Part of the U.S.G.S. Mammoth Cave
Quadrangle, Kentucky, 1955. Scale,
1:62,500; contour interval, 20 feet.

Glaciers

A *glacier* is a mass of flowing land ice derived from snow fall. The two major types of glaciers are alpine or valley glaciers, and continental glaciers or ice sheets. An *alpine glacier* is one that is confined to a valley and is literally a river of ice. An *ice sheet* or *continental glacier* covers a large surface area and is not confined to valleys. Some valley glaciers such as those found in Greenland and Antarctica, however, are fed from ice sheets, and flow in valleys that lead away from the margins of the ice sheet.

Some Principles of Glaciology

Glaciology is the study of snow and ice. Alpine or valley glaciers occur in mountain valleys where adequate snowfall sustains them. Most of this snow falls during the winter and covers the entire glacier. During the ensuing summer, some or all of the previous winter snowfall is melted and is discharged from the glacier in meltwater streams. The snow that remains at the end of the summer melt season gradually changes to ice and becomes part of the glacier.

The 12-month period of winter snow accumulation (accumulation) and summer melting (wastage) is known as the *budget year of a glacier*.[1] In either hemisphere, the budget year begins at the end of the melt season just before the first snows of the winter, and ends about 12 months later at the end of the wastage season.

During the wastage months of the budget year, the previous winter's accumulation is only partly removed from the upper reaches of the glacier. The snow that remains there lies in the *zone of accumulation*. Over the lower reaches of the glacier, the previous winter's snowfall is completely removed, and some of the underlying glacier ice is also melted. The area of the glacier that suffers wastage of both snow and ice is called the *wastage zone*. The line that separates the accumulation and wastage zones for a given budget year is the *annual snow line*. This is shown by a dashed line on the photograph of the Snow Glacier in figure 3.19 that was taken at the end of the wastage season. The white area above the snowline of the

1. The term *budget year* is synonymous with *hydrologic year*, the terminology used in some scientific literature.

Snow Glacier is the residue of the previous winter's snowfall that survived the summer melting. On the lower reaches of the Snow Glacier, all of the winter snow was melted so that bare glacier ice is exposed. Some of this ice has also been lost by summer melting. The streams flowing from the glacier terminus in the lower right-hand corner of the photograph are fed by melted snow and ice from the glacier.

It is possible to measure both wastage and accumulation for a given glacier during a budget year. The relationship between the annual wastage and accumulation over a period of years describes the general health of a glacier. For example, if wastage exceeds accumulation for a period of 10 to 20 years, the total mass of the glacier will decrease, a condition usually reflected in the retreat of the glacier terminus and a reduced thickness of the glacier. If accumulation exceeds wastage for several years, the glacier terminus advances and the glacier thickens.

A comparison between accumulation and wastage for a given budget year yields the *net mass balance* of the glacier. Net mass balance is expressed numerically in feet or meters of water equivalent, and represents a hypothetical layer of water determined by many measured thicknesses of columns of snow or ice on the glacier's surface. If a glacier gains in mass during a budget year (i.e., accumulation exceeds wastage), the glacier is said to have a *positive net mass balance*. If the glacier loses mass during the budget year, the glacier will have a *negative net mass balance*.

Glacial Geology

Glacial geology is the study of landforms and sediments produced by glaciers. A glacier erodes the surface over which it flows while at the same time functions as a conveyor belt to transport debris from the valley walls and floor to the glacier terminus where it is deposited. A terrain once occupied by valley glaciers is unique. Valley glaciers produce erosional and depositional features that are unmistakable evidence of their former existence.

Debris carried and deposited directly by a glacier is *till*, an unsorted mass of particles ranging in size from clay to boulders. Material deposited by glacially fed streams is *outwash*, which usually consists of water-sorted sands

Figure 3.19 Oblique aerial photo of Snow Glacier, Kenai Peninsula, Alaska. Annual snowline shown as dashed line. (Photograph by Austin Post, U.S.G.S.)

and gravels. Lakes fed by glacial streams contain silt and clay derived from the debris carried by the glacier. A lake in contact with a glacier terminus may contain *icebergs*, which are masses of glacier ice either floating or grounded on the lake floor.

Till accumulates in various topographic forms called moraines. Debris eroded from the walls of a valley glacier is transported along the margin of the glacier as a *lateral moraine*. The junction of lateral moraines at the confluence of two valley glaciers forms a *medial moraine* (fig. 3.20). Both lateral and medial moraines ultimately reach the glacier terminus or snout where an *end moraine* is formed.

Moraines are identified on photos of glaciers as dark bands in the wastage zone (fig. 3.19). Moraines are not visible in the accumulation zone because they are covered by the perennial snow that exists there. Lateral and medial moraines define the flow lines of a glacier and therefore cannot cross each other. An end moraine can remain long after the terminus of the glacier that formed it has retreated. Successive end moraines lying beyond the snout of a retreating valley glacier are called *recessional moraines*.

Erosional features usually dominate a terrain that was shaped by valley glaciers. A glacially eroded valley is *U-shaped* in cross-section, and its headward part may contain a *cirque*, an amphitheater-like feature. A cirque that contains a lake is called a *tarn*. *Hanging valleys* form where tributary glaciers once joined the trunk glacier, and waterfalls cascade from them to the main valley floor.

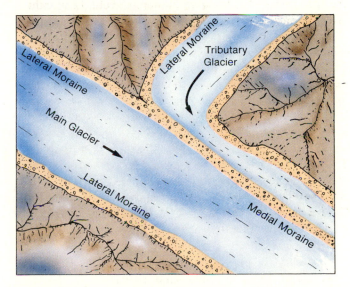

Figure 3.20 Lateral moraines join to become a medial moraine as a tributary glacier joins the main ice mass. (From Carla W. Montgomery, *Physical Geology.* Copyright © 1987 Wm. C. Brown Publishers, Dubuque, Iowa. All Rights Reserved. Reprinted by permission.)

A narrow, rugged divide between two parallel glacial valleys is an *arète*, and the divide between the headward regions of oppositely sloping glacial valleys is a *serrate divide*. A pyramidal-shaped mountain peak near the heads of valley glaciers or glaciated valleys is a *horn*, the most famous of which is the Matterhorn near Zermatt, Switzerland.

A

Figure 3.21   Oblique aerial photographs of the South Cascade Glacier, Washington. (*A*) September 27, 1960; Neg. No. FR6025-50. (*B*) October 10, 1983; Neg. No. 83R1-188. (Courtesy of Andrew G. Fountain, U.S. Geological Survey, Tacoma, Washington.)

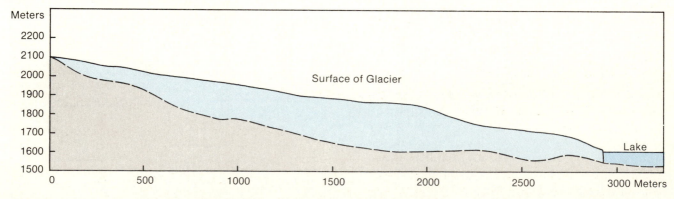

Figure 3.22   Longitudinal profile of the South Cascade Glacier during the 1965-66 budget year. No vertical exaggeration. (Based on U.S.G.S. Professional Paper 715-A, 1971.)

Annual Snow Line 1983

new snow Porosity 90%

old snow (edgemelt)

Firn First freezing of water crystals 50% Porosity

Glacial Ice 0% Porosity

B

Exercise 15. Alpine Glaciation

15A. SOUTH CASCADE GLACIER, Washington

Figure 3.21 shows two aerial photos of the South Cascade Glacier in the state of Washington. Photo *A* was taken on September 27, 1960, and photo *B* was taken on October 10, 1983. A comparison of the features shown in these two photographs reveals much about the recent history of this glacier. A longitudinal profile along the centerline of the glacier is shown in figure 3.22.

1. Draw a dashed pencil line on each of the photographs showing the boundary between the wastage and accumulation zones (ignore small patches of snow surrounded by bare ice). Label the line on photo *A* "annual snow line 1960," and the one on photo *B*, "annual snow line 1983."

2. During the time that has elapsed between the two dates of the photographs, has the snow line generally remained stationary, moved to a lower elevation, or moved to a higher elevation?

3. In comparing the two photographs, what is the evidence that the glacier has thinned in the wastage zone during the period 1960 to 1983?

4. Why are there no icebergs in the lake on photo *B*?

[*Continued on p. 100*]

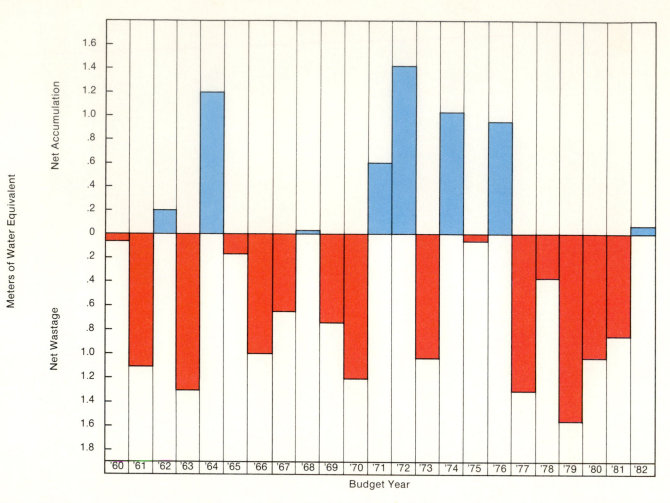

Figure 3.23 Annual net accumulation and annual net wastage on the South Cascade Glacier, Washington, for the budget years 1960 to 1982, based on data from table 3.2.

15A. *Continued*

5. The annual net mass balances of the South Cascade Glacier for each of the years 1960 through 1982 are presented in graphic form in figure 3.23, and the actual values for annual net wastage or net accumulation, expressed in meters of water equivalent over the entire glacier surface, are given in table 3.2. Determine the algebraic sum of the net balance values in table 3.2, and use the result to refute or verify the evidence of glacial retreat based on the visual inspection of the two photographs.

6. Based on the data in table 3.2, in which year did the South Cascade Glacier come closest to being in equilibrium; that is, the net mass balance was almost zero?

Table 3.2. Annual net mass balances for the South Cascade Glacier, Washington, for the budget years 1960 to 1982.

BUDGET YEAR	NET BALANCE*	BUDGET YEAR	NET BALANCE*
1960	−0.50	1972	+1.47
1961	−1.10	1973	−1.03
1962	+0.20	1974	+1.02
1963	−1.30	1975	−0.05
1964	+1.20	1976	+0.95
1965	−0.17	1977	−1.31
1966	−0.98	1978	−0.38
1967	−0.63	1979	−1.56
1968	+0.02	1980	−1.02
1969	−0.73	1981	−0.84
1970	−1.20	1982	+0.08
1971	+0.59		

Source: Courtesy of Andrew G. Fountain, U.S. Geological Survey, Tacoma, Washington.

*Net balance values expressed as meters of water equivalent spread over the entire glacier.

15B. CORDOVA MAP, Alaska

This map (fig. 3.24) shows a number of alpine glaciers. The Heney Glacier flows from the lower left-hand margin of the map area in a northeasterly direction until it ends in a lake at the upper right-hand margin of the map area. A number of small tributary glaciers feed into the Heney Glacier, and a large, three-mile long, unnamed tributary joins the Heney in section 15 about 3.5 miles south of the northern boundary of the map area. Notice that the configuration of the glacier surfaces are depicted by blue contour lines that are continuations of the brown contour lines. The C.I. and all other principles used in the interpretation of contour lines are applicable to the blue contours. The pattern of brown stippling on parts of the glaciers represents morainic material visible on the glacier surface.

1. Based on the relationship of medial and lateral moraines to the annual snow line on the Snow Glacier photograph (fig. 3.19), and assuming that the Cordova Map is based on glacier conditions at the *end* of the wastage season, determine the *maximum* elevation of the snow lines on the Heney Glacier and the McCune Glacier, and describe the reasoning used in arriving at your answer. (Assume that the snow line for each glacier is more or less coincident with a contour line across the glacier surface.)

2. Study the lower reaches of the Heney Glacier between the 1,700-foot glacier contour and the glacier terminus. Notice how the 800-, 900-, 1,000-, 1,100-, and 1,200-foot glacier contours outline a narrow ridge. Identify the axis of this ridge by a red pencil line on figure 3.24 from the 1,700-foot glacier contour to the terminus, and speculate on its origin.

3. Locate the *medial* moraine that lies between the 2,000- and 3,000-foot glacier contours on the west side of the Heney Glacier. Trace this moraine with a black pencil up-glacier to its logical point of origin. Show by black pencil the lower reaches of the two lateral moraines that formed it. Do the same for the four medial moraines above 1,500 feet on the tribuary glacier that joins the Heney at about the 1,700-foot glacier contour.

4. Locate the point in the upper reaches of the Heney Glacier where the 4,500-foot contour line intersects the narrow rock outcrop south of the word *crevasses*. Assume that a large boulder becomes dislodged from the eastern edge of this rock island and falls onto the glacier surface. Using a blue pencil, draw a line on figure 3.24 from the point where this boulder begins its journey down-glacier to the point where it reaches the glacier terminus. (This line is a *flow* line and cannot cross a medial moraine.) If the average velocity of the glacier is 2 feet per day, and assuming that the glacier terminus remains fixed for the entire period, how long will it take for the glacier to transport the boulder along the route defined by your pencil line?

5. Locate two peaks, one on either side of the Heney Glacier below its junction with the western tributary. The northern of the two peaks is marked by a brown cross labeled with the black numbers 4,785. The southern peak is marked with a brown cross labeled with the black numbers 4,144. Draw a light pencil line on figure 3.24 connecting the two peaks. At the point where this line intersects the medial moraine lying .2 miles from the southern margin of the glacier, assume that a prospector discovered a boulder containing a high concentration of gold. Outline with red pencil the area on the map wherein the prospector should begin an investigation to find the *bedrock* source of this boulder.

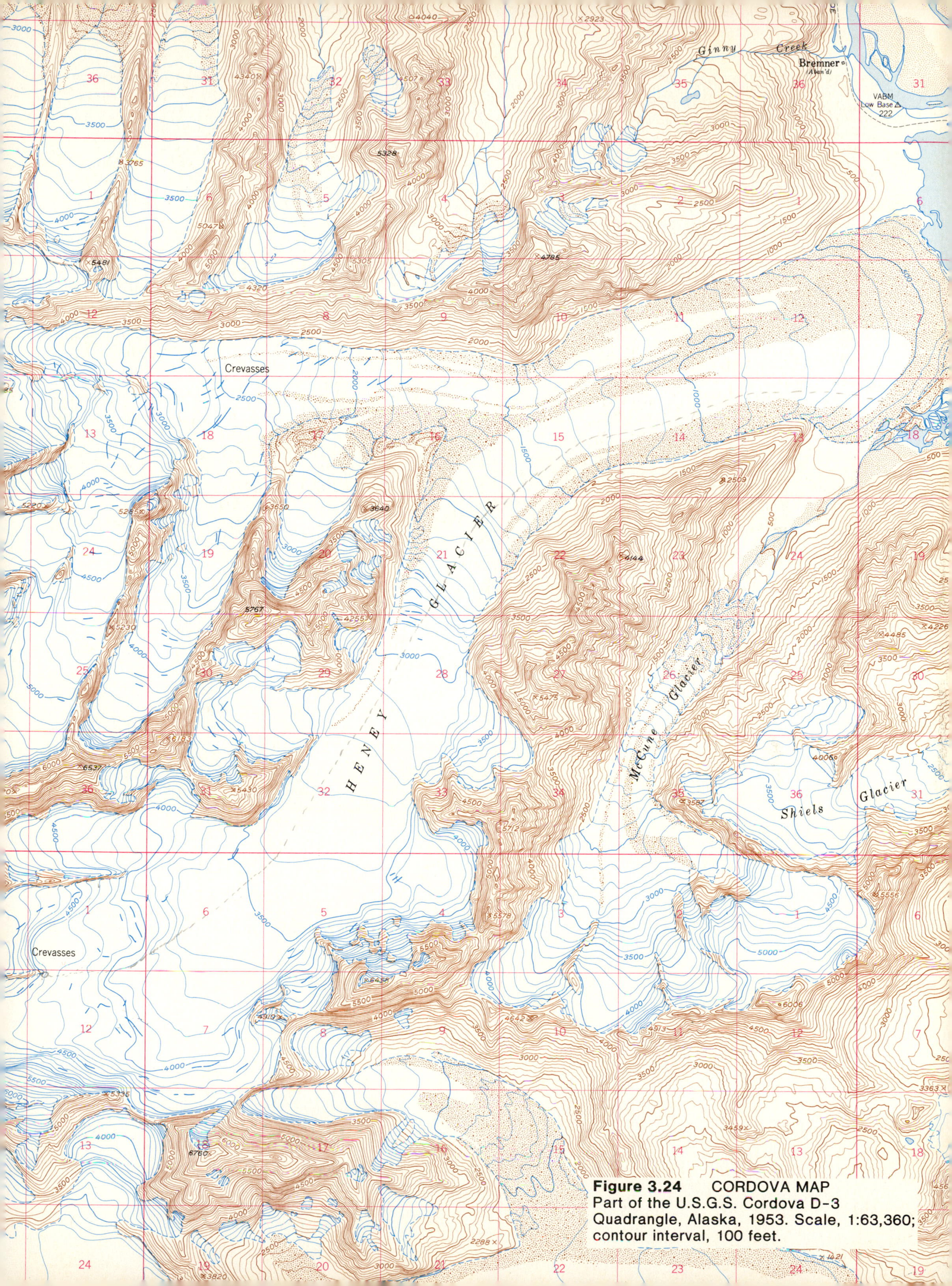

Figure 3.24   CORDOVA MAP
Part of the U.S.G.S. Cordova D-3
Quadrangle, Alaska, 1953. Scale, 1:63,360;
contour interval, 100 feet.

15C. MATTERHORN PEAK MAP, California

The area shown in the Matterhorn Peak Map (fig. 3.25) was once occupied by a system of valley glaciers. The valleys produced by these glaciers now contain streams whose courses more or less coincide with the longitudinal axes of the former glaciers.

1. Trace the existing drainage lines in blue pencil. How do the valleys containing these streams differ in cross-section from the stream valleys south of the Mogollon Rim on the Promontory Butte Map of figure 3.3?

2. Sawtooth Ridge on the Matterhorn Map is the drainage divide between south-flowing streams such as Rock Creek and Spiller Creek, and streams flowing north to Robinson Creek.
   a) Visualize a topographic profile along the dashed line that defines the crest of Sawtooth Ridge. Does this profile lie more or less parallel to the contour lines or does it generally cut across them?
   b) Visualize a profile along the dashed line that approximates the divide on the Promontory Butte Map of figure 3.3. Does this profile more or less parallel the contour lines or does it generally cut across them?
   c) Why do the two profiles on the two map areas differ?

3. Assign an appropriate physiographic name to each of the following features shown on the Matterhorn Map:
   a) Stanton Peak (SE quadrant of the map).
   b) Spiller Lake (SW quadrant of the map).
   c) The part of Horse Creek Valley lying up-valley from the waterfalls at about the 7,600-foot contour line (NE quadrant of the map).
   d) The Cleaver (central part of the map).
   e) Sawtooth Ridge.

4. The topography of the Matterhorn Peak Map area is dominated by landforms produced by glacial erosion. During the time when the former valley glaciers were receding, it is not unreasonable to assume that recessional moraines might have been formed. Remnants of an end moraine are represented by the 7,120-foot contour line north of the Twin Lakes Campground in the extreme northeast corner of the map. With this one exception, no other recessional moraines are in evidence on the entire map. Suggest a reason for this.

Figure 3.25   MATTERHORN PEAK MAP
Part of the U.S.G.S. Matterhorn Peak
Quadrangle, California, 1956. Scale,
1:62,500; Contour interval, 80 feet.

Exercise 16. Continental Glaciation

16A. STEREOPAIR OF DRUMLINS, Wisconsin
PALMYRA MAP, New York

In many areas formerly covered by continental glaciers, deposits of till dominate the landscape. Under certain circumstances the till was moulded by the moving ice into swarms of elliptical hills called *drumlins*. A drumlin's long axis is roughly parallel to the direction of flow of the ice. The longitudinal profile of a drumlin characteristically shows a steep side on the end from which the ice was coming, and a more gentle slope on the downstream direction of ice flow.

Figure 3.26 is a stereopair of part of a drumlin field in Wisconsin. Agricultural fields developed on drumlins exhibit a technique of "contour plowing," a practice that impedes surface run-off and reduces soil erosion. The boundaries of the ploughed fields are more or less parallel to random contour lines on a map, and help in identifying the shapes of these features on aerial photographs.

1. Draw arrows along the axes of the drumlins in figure 3.26 showing the direction of ice movement.

[*Continued on p. 108*]

Figure 3.26 Aerial photograph stereopair of drumlins in Wisconsin. Scale, 1:26,000. (Courtesy of U.S.G.S.)

16A. *Continued*

   On figure 3.27, two geologic processes were instrumental in shaping the landscape depicted by the contours on the map. Continental glaciation produced the dominant topography, and the stream system is a post-glacial phenomenon. The following questions relate to the Palmyra Map of figure 3.27.

2. What is the dominant glacial landform in this area?

3. Draw a pencil line along the longitudinal profile of Baker Hill. Describe the difference between the north and south ends of this profile.

4. Based on the description of the shape of the longitudinal profile of Baker Hill, draw an arrow point on the appropriate end of the profile line showing the direction of ice movement.

5. In what way has the location of the North Central Railroad tracks been influenced by the elongate hills through which they pass?

6. To what extent, if any, have the existing streams modified the topography of this area?

Figure 3.27 PALMYRA MAP
Part of the U.S.G.S. Palmyra Quadrangle,
New York, 1932. Scale, 1:62,500; contour
interval, 20 feet.

16B. PASSADUMKEAG MAP, Maine

Two north-south trending eskers are well-displayed on the map in figure 3.29. Eskers are believed to be produced by the deposition of sand and gravel in ice tunnels associated with continental glaciers. An alternative possibility is that eskers are formed as the deposits in meltwater ice channels on the surface of a continental glacier. Water flowing in an ice tunnel has the same mode of flow as water flowing in a large diameter pipe or conduit, whereas water flowing in an ice-surface channel flows in the same way that water moves in a normal river channel. Figure 3.28 is a stereopair of an esker in Michigan for reference study.

1. Draw the approximate boundaries of the two eskers on the map. Use a brown pencil. What local name is given to these eskers?

2. Give the elevations and locations of the highest and lowest points on the esker that lies to the east of the Penobscot River. Describe in general terms the longitudinal profiles of the two eskers.

3. What map evidence reveals the nature of the geological materials contained in or associated with the eskers?

4. Considering the shape and form of an esker, and the presence of gravel in eskers, explain why it is necessary to invoke the presence of glacier ice to account for their origin.

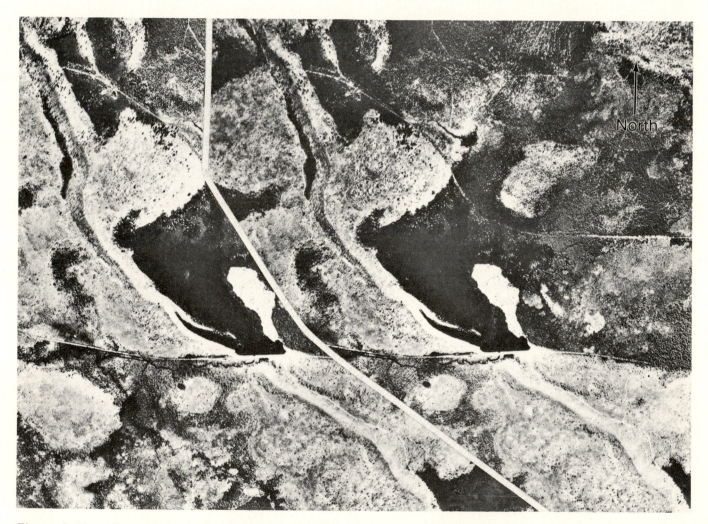

Figure 3.28   Stereopair of part of an esker in Michigan. Scale, 1:24,000. (Courtesy of U.S. Geologic Survey.)

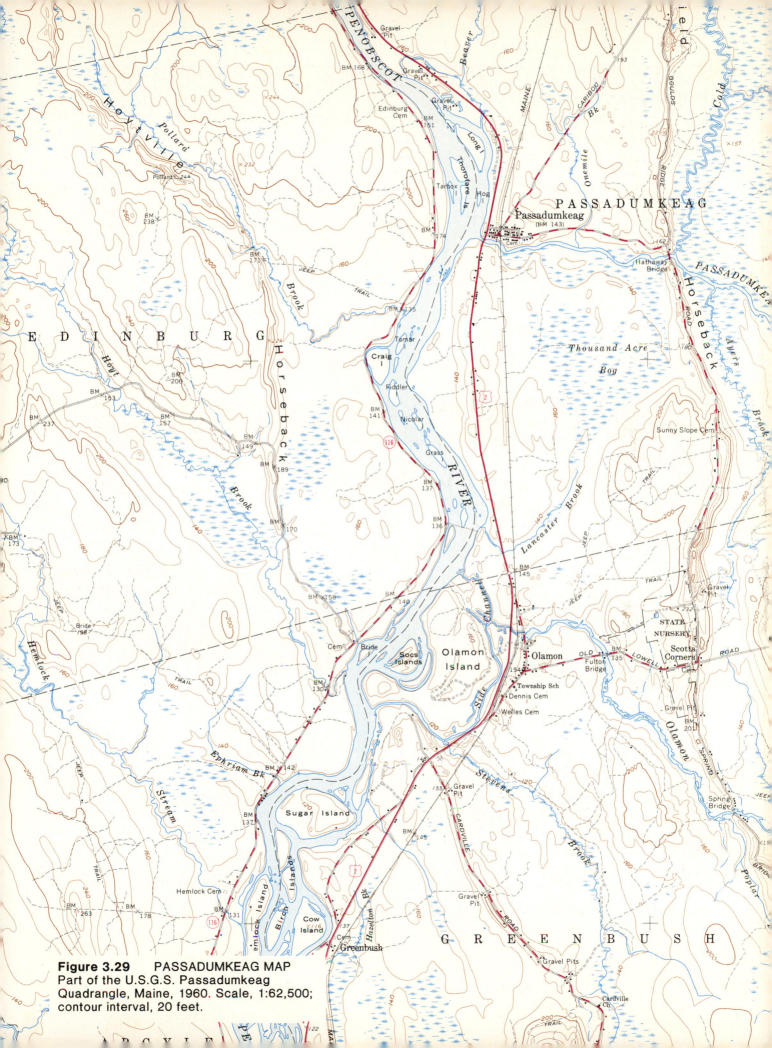

Figure 3.29   PASSADUMKEAG MAP
Part of the U.S.G.S. Passadumkeag
Quadrangle, Maine, 1960. Scale, 1:62,500;
contour interval, 20 feet.

16C. WHITEWATER MAP, Wisconsin

The topography of this map area (fig. 3.32) was produced by the last advance and retreat of the continental ice sheet. Three general terrain types are present: (1) ground moraine, (2) end moraine, and (3) pitted outwash plain. Because the area covered by the map is so small compared to the total area covered by the continental glacier, the relationship of the Kettle moraine (an end moraine) to the position of the ice front at the time the moraine was built is not immediately clear from a casual inspection of the map.

Let us assume that the Kettle moraine on the Whitewater Map is part of an end moraine that was deposited by the "Kettle lobe"* as shown in the sketch map of figure 3.30. On that sketch map, two possible locations of the Whitewater Map are shown as *A* and *B*. Each shows the Kettle moraine trending across the map area from a southwesterly to a northeasterly direction, as it does on the Whitewater Map. Beyond the end moraine of figure 3.30, outwash associated with the "Kettle lobe" is shown. In addition, we can assume that the area formerly underlain by the "Kettle lobe" will be covered with ground moraine.

1. Study the Whitewater Map and the topographic profile of figure 3.31 and determine which of the two locations, *A* or *B* on figure 3.30 is the one that most likely represents the Whitewater Map. State in concise terms the reasons for your choice.

2. Draw the ice-surface *profile* of the "Kettle lobe" on figure 3.31 as it may have been when the "Kettle lobe" was building the Kettle moraine. (In drawing the ice-surface profile of the "Kettle lobe," assume that the ice slopes upward from the highest point on the end moraine until it attains a thickness of about 500 feet in a horizontal distance of about 4 miles.)

*A lobe is a lobate appendage of a continental glacier. Much of the glacial deposits of the Great Lakes Region of North America were deposited by lobes of the ice sheet that covered the area during parts of "the Great Ice Age" or Pleistocene Epoch as it is technically known. These lobes ranged in breadth from less than 50 to more than 100 miles.

Figure 3.30   Map of the hypothetical "Kettle lobe" showing two possible locations of the Whitewater Map (fig. 3.32), *A* or *B*. (*See* Exercise 16C for instructions.)

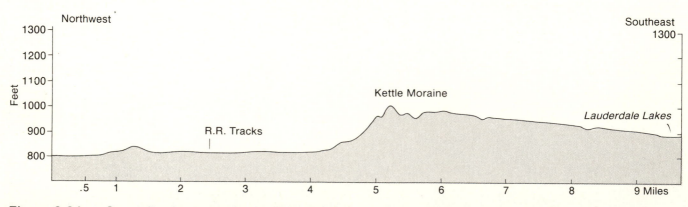

Figure 3.31   Generalized topographic profile from northwest to southeast across the Kettle moraine of the Whitewater Map (fig. 3.32). (Vertical scale exaggerated.)

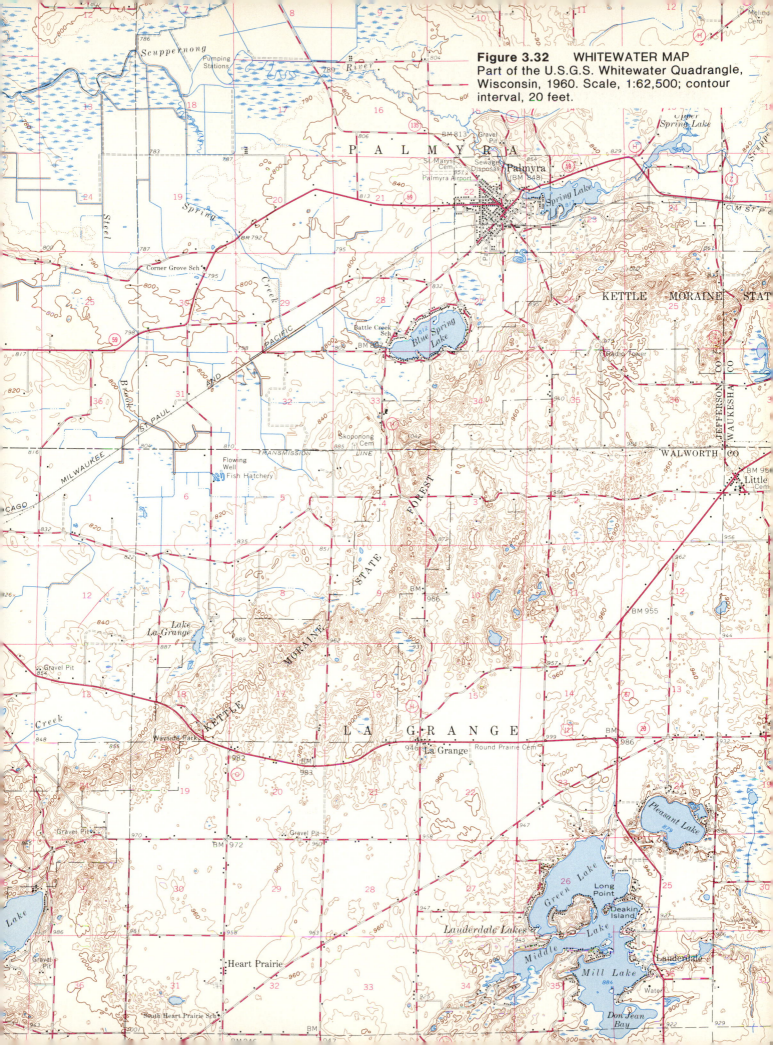

Figure 3.32   WHITEWATER MAP
Part of the U.S.G.S. Whitewater Quadrangle,
Wisconsin, 1960. Scale, 1:62,500; contour
interval, 20 feet.

Exercise 17. Landforms Produced by Wind Action

17A. KANE SPRING BARCHANS, California
KANE SPRING STEREOPAIR, California

These crescent-shaped sand dunes (fig. 3.33) are active barchans. They are moving across the surface of what once was the floor of the Salton Sea in Southern California. Where not occupied by sand dunes, the land surface supports very little vegetation, and contains extensive tracts of lag gravel. Measurements on the change in positions of the barchans show that they moved 325 to 925 feet during a 7-year period. Study the stereopair in figure 3.34 and compare it with the topographic map of figure 3.33. Note the difference in scales between the air photos and the topographic map.

1. What is the direction of the prevailing wind in this area? How determined?

2. Assuming that an airstrip should be aligned, more or less, with the prevailing wind direction, is the landing strip properly aligned?

3. The shapes of the dunes and the configuration of the dune pattern in Sections 25 and 30 are not as clearly shown on the topographic map (fig. 3.33) as they are on the aerial photographs of the stereopair (fig. 3.34). Some small dunes that are clearly visible on the stereopair are not shown on the topographic map. One of these is about 1,580 feet north of BM—109 (SW corner of Sec. 29). Why does this feature not appear on the topographic map when it is so clearly visible on the aerial photograph?

4. Assuming that the dunes will continue their general rate and direction of migration within the maximum and minimum limits already stated, what will be the shortest possible and longest possible times (in years) for the barchan about ½ mile west of the west end of the landing strip to reach the landing strip?

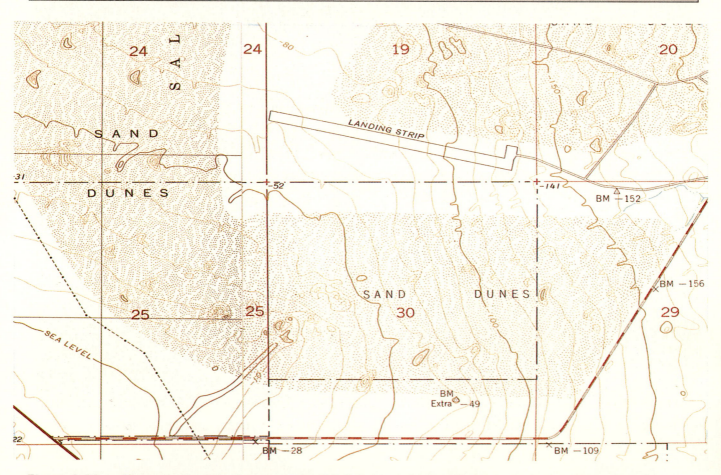

Figure 3.33   Topographic map of parts of the U.S.G.S. Kane Spring NE and Kane Spring NW quadrangles, California, 1956. Scale, 1:24,000; contour interval, 10 feet.

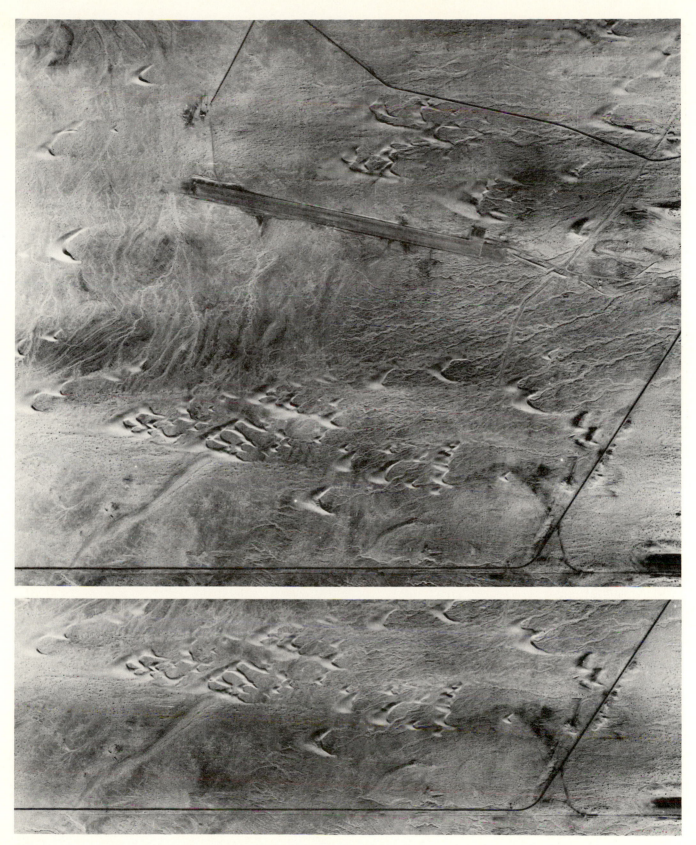

Figure 3.34 Stereopair of Kane Spring sand dunes, California. Scale, 1:20,000; date flown, November 10, 1959.

17B. LANDSAT IMAGE, Sand Hills, Nebraska ASHBY MAP, Nebraska

Figure 3.35 is part of a Landsat image and is similar to a black-and-white aerial photograph. It was taken from an altitude of 570 miles on January 9, 1973, over western Nebraska when the area was blanketed with snow. The snow cover enhances the definition of topographic features. On this image, the northern three-fourths of the area is part of the Nebraska Sand Hills. The topography is dominated by hills of wind-deposited sand that were formed at a time when the rainfall of the area was less than that at present. These dunes are now stabilized by natural grass which, along with abundant water, provides ideal conditions for cattle grazing. The east-west trending darker lines in the southern part of the image are the North and South Platte Rivers, which converge at the town of North Platte, Nebraska (near the eastern margin of the image) to form the North Platte River that flows easterly across the state of Nebraska to join the Missouri River at the Iowa-Nebraska border. The light gray elongate feature along the North Platte River is Lake McConaughy (with its winter cover of snow and ice) formed by an earth-fill dam at its eastern end. The rectangular area in the northwestern part of the image is the approximate area covered by the Ashby Map (fig. 3.14).

1. Determine the scale of the Landsat image in miles per inch and the R. F.

2. Compare the Landsat image with the Ashby Map. Which natural or man-made features shown on the Ashby Map are identifiable on the Landsat image? Which features shown on the Ashby Map are not visible on the Landsat image? Why not?

3. Draw several short arrows (¼ inch) in red pencil on the Landsat image showing the wind direction at the time of dune formation. How deduced?

4. The Landsat image shows a rectangular pattern south of the South Platte River. What is the most likely cause of this pattern?

Reference

Keech, C. F., and Bentall, R. 1971. *Dunes on the plains, the sand hills region of Nebraska.* Resources Report No. 4, Univ. Nebraska, Conservation and Survey Division, Lincoln, Nebraska, 18 pp.

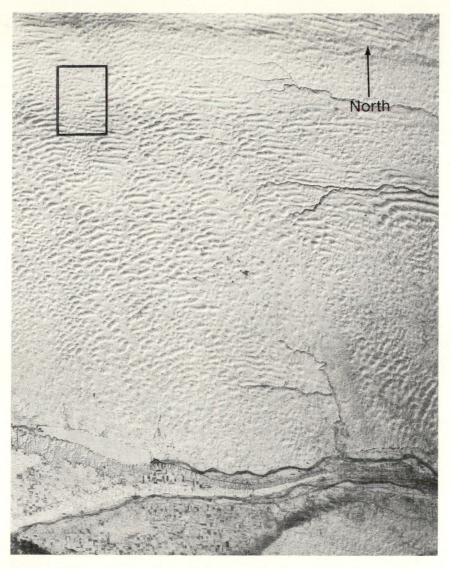

Figure 3.35 Landsat image of part of Western Nebraska made on January 9, 1973 from ERTS-1 at an altitude of 570 miles. The rectangle in the northwest corner is the area covered by the Ashby Map of figure 3.14. (NASA ERTS E-1170-17020, Courtesy of NASA and the U.S.G.S. EROS Data Center, Sioux Falls, South Dakota 57198.)

Exercise 18. Modern and Ancient Shorelines

18A. POINT REYES MAP, California

This map (fig. 3.36) shows part of the California coast near San Francisco. (North is toward the bound margin of the map.)

1. What is the evidence that the sand of Point Reyes Beach has been reworked by the wind? Draw short red arrows at a number of locations along Point Reyes Beach to show the direction of prevailing winds.

2. Draw dashed red arrows just offshore along Point Reyes Beach to show the direction of longshore drift produced by the prevailing winds.

3. The configuration of the sea floor in Drakes Bay and along the rest of the coastal area is shown by blue contour lines to a depth of 60 feet. Explain why the 60-foot depth contour is much closer to the shoreline off Point Reyes than it is in Drakes Bay.

4. Draw the position of the shoreline on the Point Reyes Map as it would appear if sea level were raised 80 feet. Use a red pencil.

5. Draw the position of the shoreline on the Point Reyes Map as it would appear if sea level were lowered 30 feet. Use a blue pencil.

6. The embayment formed by Barries Bay, Creamery Bay, Schooner Bay, Home Bay, and Drakes Estero is very irregular in outline. Explain the reason for this.

7. The west shore of Drakes Bay appears on the map as a smooth curve. Use a black pencil to reconstruct this shoreline between the mouth of Drakes Estero and Point Reyes as it might have appeared at an earlier time in its history, but *after* any significant changes in sea level occurred.

8. The coastal side of Point Reyes has an entirely different physiographic character than either Point Reyes Beach or the shoreline of Drakes Bay. Describe the character of the Point Reyes coastline and explain why it is so different from the other two.

Figure 3.36 POINT REYES MAP
Part of the U.S.G.S. Point Reyes Quadrangle, California, 1954. Scale, 1:62,500; contour interval, 80 feet.

18B. THE MISSISSIPPI DELTA, Louisiana

The Mississippi River discharges into the Gulf of Mexico *via* the Mississippi Delta, which has been formed from sediment carried by the river over thousands of years. The delta is actually a composite of several subdeltas, each of which was active during a phase of the growth of the larger feature. Each subdelta consists of a system of distributary channels called *passes*.

Figure 3.38 shows different courses of the Mississippi River during the last 2,000 years. Four distinct subdeltas are shown, and each is related to a change in the course of the river since about A.D. 100. The oldest of the four subdeltas (the Teche Delta) was active for about 200 to 300 years when the ancestral Mississippi River occupied a position (course 1 in fig. 3.38) 40 to 50 miles west of its present course south of Vicksburg. The Present Delta was established by a natural diversion of the river near New Orleans about A.D. 1500–1600.

Figure 3.37 is a false color image of part of the Mississippi Delta showing parts of the four subdeltas identified in figure 3.38. The Present Delta is clearly visible with its plume of suspended sediment delivered to the Gulf of Mexico by the active distributary system. Study figures 3.37 and 3.38, paying special attention to the physiographic differences between the subdeltas. For each subdelta, notice in particular the distributaries, coastline, and any attending offshore features.

1. Write a discussion of the evolution of a subdelta from the time it first becomes established as an active discharge point of the river until long after it ceases to function in that capacity. Where possible, use examples of features associated with the various subdeltas identified in figure 3.38. Divide your discussion into three parts, each of which represents a different stage in the life history of a subdelta. Stage I is represented by the subdelta associated with the modern course of the river. Stage II is reflected in the features of the St. Bernard Delta, and Stage III is represented by the combined Teche and Lafourche Deltas. Give some indication of the time, in years, covered by each stage. Use data from figure 3.38.

[*Continued on p. 121*]

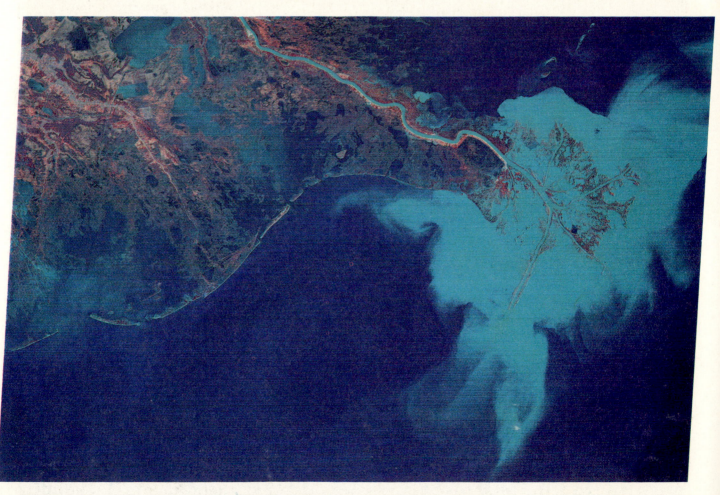

Figure 3.37   False color image of the Mississippi Delta made from Landsat on April 3, 1976. (Courtesy of NASA and the U.S.G.S. EROS Data Center, Sioux Falls, South Dakota 57198.)

18B. *Continued*

2. In the middle of this century, the Mississippi River discharged about 25% of its volume into the Gulf of Mexico *via* the Atchafalaya River (fig. 3.38). This percentage would have increased ultimately to a point at which a natural diversion of the entire Mississippi through the Atchafalaya River would have occurred. This natural event was forestalled in 1963 when the U.S. Corps of Engineers completed a control dam at the potential diversion site. Describe what the impact on the areas south of the diversion might have been had not this human intervention to a potential natural geologic event occurred.

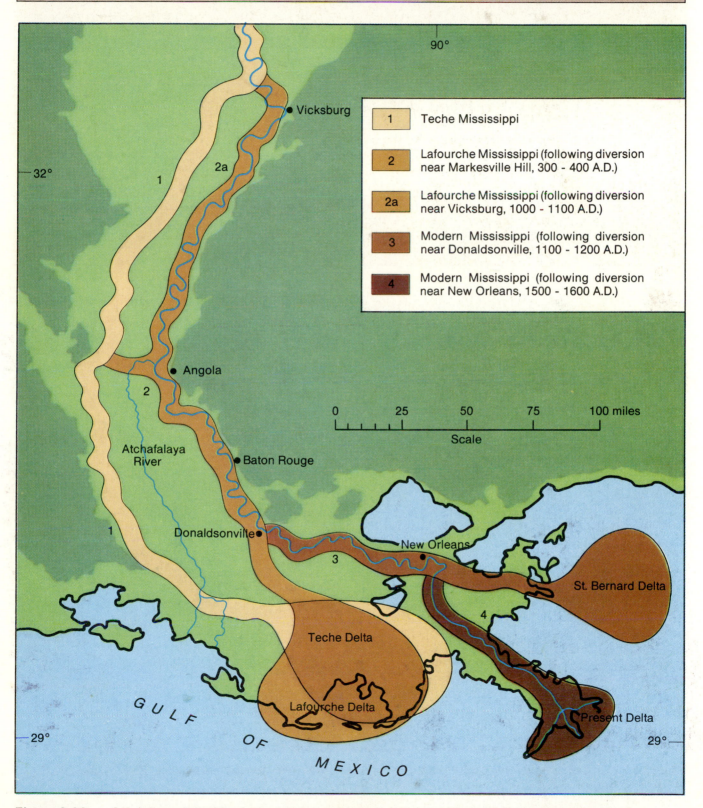

Figure 3.38 Subdeltas of the Mississippi Delta and associated courses of the ancestral and modern Mississippi River. (From W. D. Thornbury, *Regional Geomorphology of the United States.* New York: John Wiley and Sons, Inc., 1956, p. 61.)

18C. HATTERAS INLET, North Carolina

Hatteras Inlet is a break in one of the long sandy islands that fringe the Atlantic coast of the southeastern United States. The zone between the long sandy islands and the mainland is called a *lagoon*. Lagoons are connected to the open sea by inlets. Hatteras Island is one of the islands that fringes the coast of North Carolina, and its southwestern tip is shown on the aerial photos in figure 3.40, *A* and *B*. Hatteras Inlet lies to the west of Inlet Peninsula (fig. 3.39). Photograph *A* was taken on March 29, 1955, and *B* was taken on August 16, 1959, about one year after Hurricane Helene struck the Atlantic coast in September 1958. The south shore of Hatteras Island faces the Atlantic Ocean and the north shore is toward the lagoon.

Inlet Peninsula (fig. 3.39) is a recurved spit formed by the waves and currents prevailing along this coast most of the time. Hurricane winds generate high waves and strong tidal currents that cause profound changes along the coastlines.

1. Using the map of figure 3.39 as a base, sketch the shape of the Inlet Peninsula as it appears in figure 3.40*B*. (Note that the scales of the map and the photograph are roughly the same.) What major changes were wrought by the hurricane?

2. Considerable sand appears on the shallow lagoonal floor in figure 3.40*B*. This sand does not appear in figure 3.40*A*. What is the probable source of this sand?

3. Use a red pencil to trace the crest of the successive growth stages of Inlet Peninsula on figure 3.40*A*.

4. What evidence is visible in figure 3.40*B* indicating that Inlet Peninsula may be reforming?

5. What is the evidence that, in spite of the dramatic changes produced by hurricanes, Hatteras Island has experienced "permanent" westward growth?

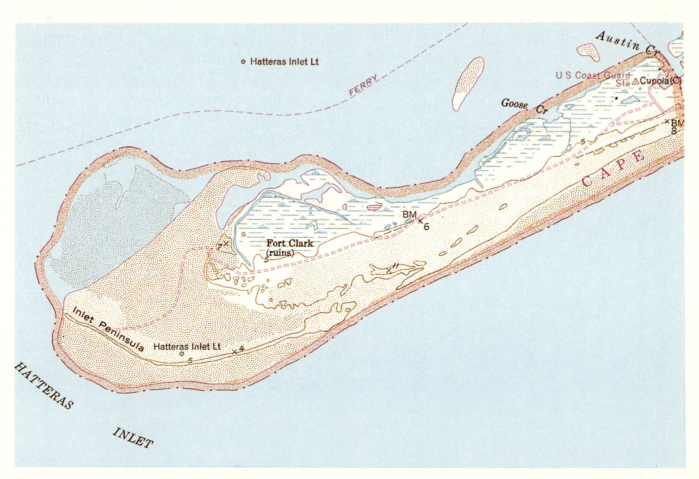

Figure 3.39   Part of the U.S.G.S. Hatteras Inlet Quadrangle, North Carolina, 1948. Scale, 1:24,000; contour interval, 5 feet.

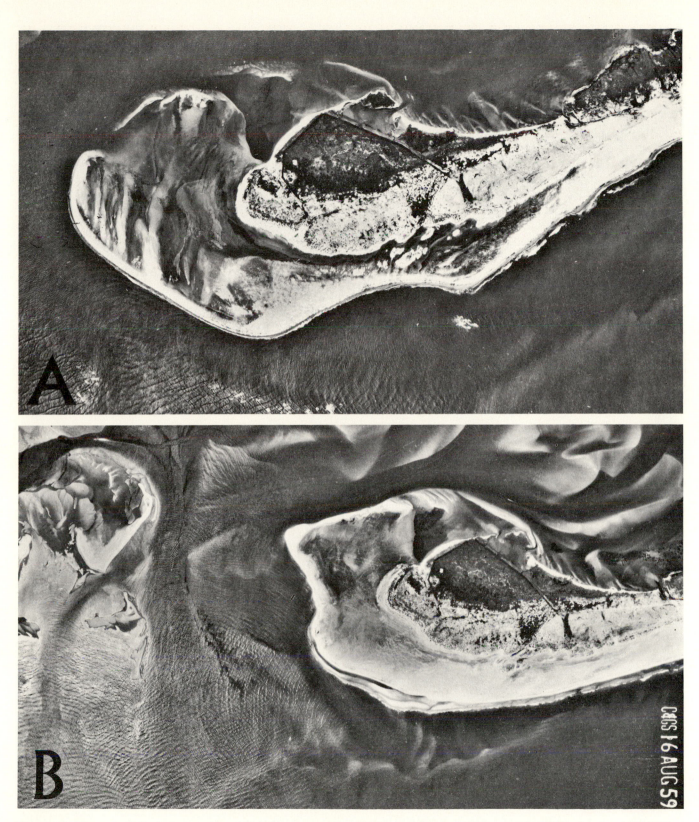

Figure 3.40 Aerial photographs of the western end of Hatteras Island. *(A)* March 29, 1955. Scale, 1:20,000. *(B)* August 16, 1959. Scale, 1:24,000. Hurricane Helene struck this area in September 1958. (This is *not* a stereopair.)

18D. NORTH OLMSTEAD MAP, Ohio

This map (fig. 3.42) shows an area along the south shore of Lake Erie. Detroit Road and Center Ridge Road follow two shorelines formed by ancestral stages of Lake Erie. These will be referred to as the Detroit shoreline and the Center Ridge shoreline, respectively.

Shorelines are generally of two types, erosional or depositional. An erosional shoreline is one characterized by a *wave-cut* cliff, and a depositional shoreline is characterized by *beach ridges*. The stretch of the modern shoreline of Lake Erie shown on the North Olmstead map lies at the base of a typical wave-cut cliff.

Figure 3.41 is a north-south profile that crosses part of the North Olmstead Map. The profile begins at the 30-foot depth contour in Lake Erie about .7 mile west of the east margin of the map. From there, the line of profile extends due south as a straight line along Forest View Road, and continues south as it crosses Wolf Road, the railroad tracks, Detroit Road, and Center Ridge Road. The south end of the profile is at the point where the above defined line intersects the 740-foot contour, about halfway between Center Ridge Road and Westwood Avenue.

1. With black pencil, draw the line of profile on the North Olmstead Map. With reference to the profile (fig. 3.41), compare the Detroit and Center Ridge shorelines with the modern shore of Lake Erie. Which shorelines are erosional? Which depositional?

2. Use a blue pencil and a straightedge to draw (on the profile) the water surfaces of the lakes that produced the Detroit and Center Ridge shorelines. Give the approximate elevation of each.

3. Draw a solid line on the topographic profile of figure 3.41 between Detroit Road and Wolf Road to show what the topographic profile might have been prior to the establishment of the Detroit shoreline. Show, by a red dashed line, a profile intermediate between the one just drawn and the present profile.

4. The Detroit and Center Ridge shorelines are related to separate water levels associated with the precursors of modern Lake Erie. Which of the two premodern shorelines on the North Olmstead Map is the oldest? How determined, and what assumptions were made?

5. Examine the map carefully, especially the lower right quadrant. Locate and describe any evidence that suggests a water level *higher* than the one associated with the Center Ridge shoreline.

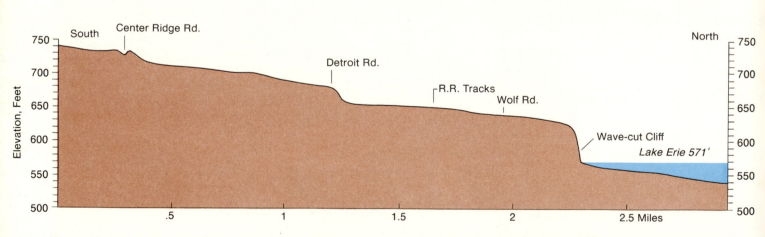

Figure 3.41 North-south topographic profile based on the North Olmstead Map (fig. 3.42) showing some ancient shorelines related to the ancestral stages of Lake Erie, Ohio. (*See* Exercise 18C for exact location of the profile.)

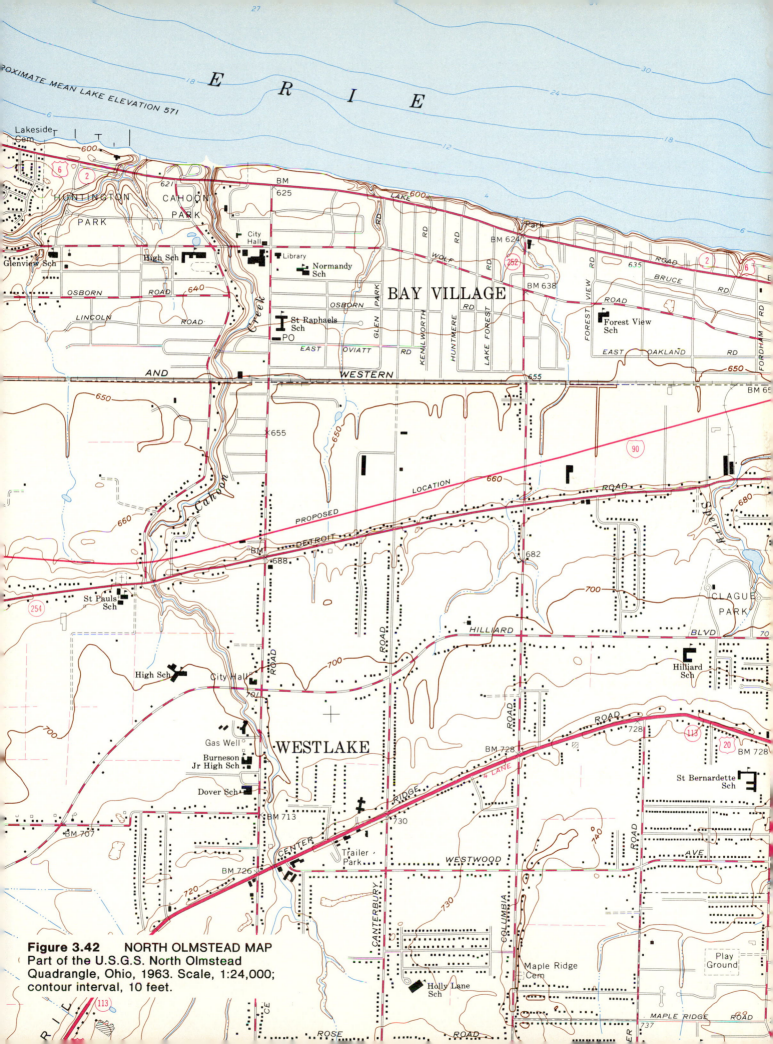

Figure 3.42   NORTH OLMSTEAD MAP
Part of the U.S.G.S. North Olmstead
Quadrangle, Ohio, 1963. Scale, 1:24,000;
contour interval, 10 feet.

18E. GREAT LAKES, Michigan

The Great Lakes straddle the U.S.-Canadian border and have a combined surface area of almost 100,000 square miles (fig. 3.43). The lakes came into being during the retreat of the Pleistocene ice sheet that once covered the entire area more than 10,000 years ago. The volume of water in the modern lakes is mainly a function of surface runoff from the surrounding drainage basin, evaporation from the lakes themselves, and the outflow of water to the Atlantic Ocean via the St. Lawrence River. The volume of water stored in the Great Lakes is reflected in their water levels. These levels are published monthly for each lake by the U.S. Army Corps of Engineers (fig. 3.44) and are a matter of public record.

The shorelines of the Great Lakes are sites of thousands of vacation homes and permanent dwellings. Those with lake frontage are desirable from an aesthetic point of view, but they are in considerable jeopardy during periods of high water levels when storm waves destroy sandy beaches formed during low water stages, and cause severe recession of the bluffs and cliffs by wave erosion at their bases.

Much of the eastern shoreline of Lake Michigan is characterized by steep bluffs that are formed in old sand dunes and other unindurated Pleistocene sediments. Some of the bluffs rise to more than 100 feet above the lake. Private dwellings on the tops of these bluffs were built during periods of low water when the shore between the bases of the cliffs and the water's edge were characterized by wide sandy beaches. Those who purchased or constructed homes during low water stages believed that the wide beaches fronting their properties were a permanent part of the landscape and provided adequate protection from any future wave erosion. These were false assumptions as many who owned shore property on the eastern shore of Lake Michigan learned at great cost during the period 1950 through 1987.

Three times during this period, 1952–1953, 1972–1976, and 1985–1987, the water levels of Lake Michigan stood at extraordinarily high levels, and twice during the same period, 1958–1959 and 1964, the levels were extremely low (fig. 3.44). The water level of 581.6 feet in 1986 was the highest on record since 1900, and the water level of 575.7 feet in 1964 was the lowest on record for the same period. Thus, in the 22-year period between 1964 and 1986, the level of the lake varied by about 6 feet, mainly by natural causes.

The damage to property during high water stages is all too apparent in the photographs of figures 3.45 and 3.46. The house in figure 3.45 was on the brink of disaster when it was photographed in 1973, and the one in figure 3.46 was abandoned by the time it was photographed in 1986.

One might ask why these houses and hundreds of others like them were built in the first place. The answer lies in the lack of understanding of the relatively short time it takes for the lake level to change drastically, and the inability of anyone to predict future levels over a time period of a decade or so. No one would think of building a house next to the two in figures 3.45 and 3.46 today. The consequences of such folly are all too apparent. But, when the houses along Lake Michigan and other Great Lake shores were built during low water stages, they seemed secure from the damage and destruction to which they were subjected in later years.

The lesson to be learned from this is that those who contemplate purchasing or building homes should be aware of geologic hazards, and endeavor to acquire as much information as possible about building sites from the public record before proceeding.

1. The sediments exposed in the wave-cut cliff of figure 3.45 are very susceptible to wave erosion at the cliff base by storm waves. If this material had been unweathered granite, would the house built there be any more secure than it appeared to be in 1973?

2. What evidence is there in figure 3.45 that an attempt was made to impede if not stop the erosive action of storm waves?

3. The cliff in figure 3.45 had receded tens of feet between the time the house was built in the 1930s and the date of the photograph. What happened to the material eroded from the cliff face during this period?

4. Describe the characteristics of the cliff in figure 3.45 as it might have appeared to an observer walking along its base during the 1930s when lake levels remained below the mean water level of 578.7 feet for an entire decade.

5. What must happen to the level of Lake Michigan before the stability of cliffs and bluffs along its shore can be restored?

6. The abandoned house perched on top of an ancient sand dune in figure 3.46 appears doomed to failure. Why did the owner apparently decide to abandon it rather than have it moved landward beyond immediate danger?

7. The face of the sand bluff in figure 3.46 is littered with debris derived from the house above. Aside from this, what other signs are visible on the face of the bluff that indicate it was undergoing rapid erosion at the time the photograph was taken?

Reference

National Geographic, Vol. 172, No. 1, July 1987. (A popular account of the nydrology of the Great Lakes.)

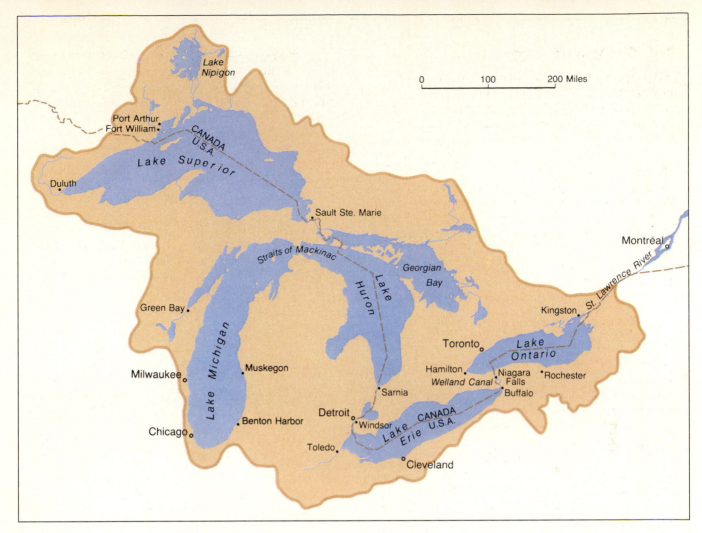

Figure 3.43 Map of the Great Lakes and the land area supplying runoff to them. The five lakes have a combined surface area of almost 100,000 square miles, and the drainage basin containing them is roughly twice that size.

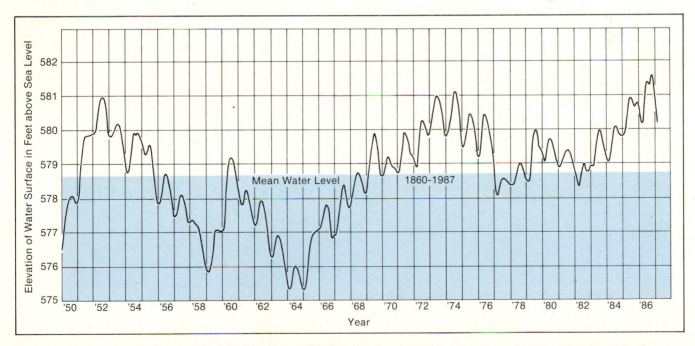

Figure 3.44 Water levels of Lake Michigan from 1950 to 1986. (Based on data published by the U.S. Army Corps of Engineers, Detroit, Michigan 48231.)

Figure 3.45   Photograph of a part of the Lake Michigan shoreline near Benton Harbor, Michigan, taken April 16, 1973. (Hann Photo Service, Hartford, Michigan.)

Figure 3.46   An ancient sand dune exposed to wave erosion on the shore of Lake Michigan near Muskegon, Michigan taken in November 1986. (Photo by Marge Beaver, Muskegon, Michigan.)

Exercise 19. Ancient and Modern Volcanoes

19A. GLASS MOUNTAIN, California

The stereopair (fig. 3.47) shows lava flows of obsidian that form Glass Mountain. Two separate episodes of extrusion are represented by two distinct lava flows.

1. Draw the boundaries of the two flows in pencil on the right-hand photo of the stereopair as you view the photos in stereovision. When you are reasonably certain that these boundaries are correct, trace the pencil lines in red for the boundary of the younger (upper) flow and green for the older (lower) flow. Shade the area of each flow on the photograph lightly with the respective colors.

2. The surface textures of these two flows are similar in appearance. Both show very little signs of erosion, yet they are distinct flows. Does this similarity suggest that the two flows are separated by a long period of geologic time, or that they are the results of two episodes of extrusion separated by a relatively short interval of geologic time? Explain briefly.

3. Part of a lava flow older than the two identified in question 1 is visible on the Glass Mountain stereopair. On the right-hand photo of the stereopair, draw the visible boundary of this older flow with a brown pencil. Based on the surface texture of this flow, does it appear to have come from the same source as the two younger flows or from some other source? Explain your answer.

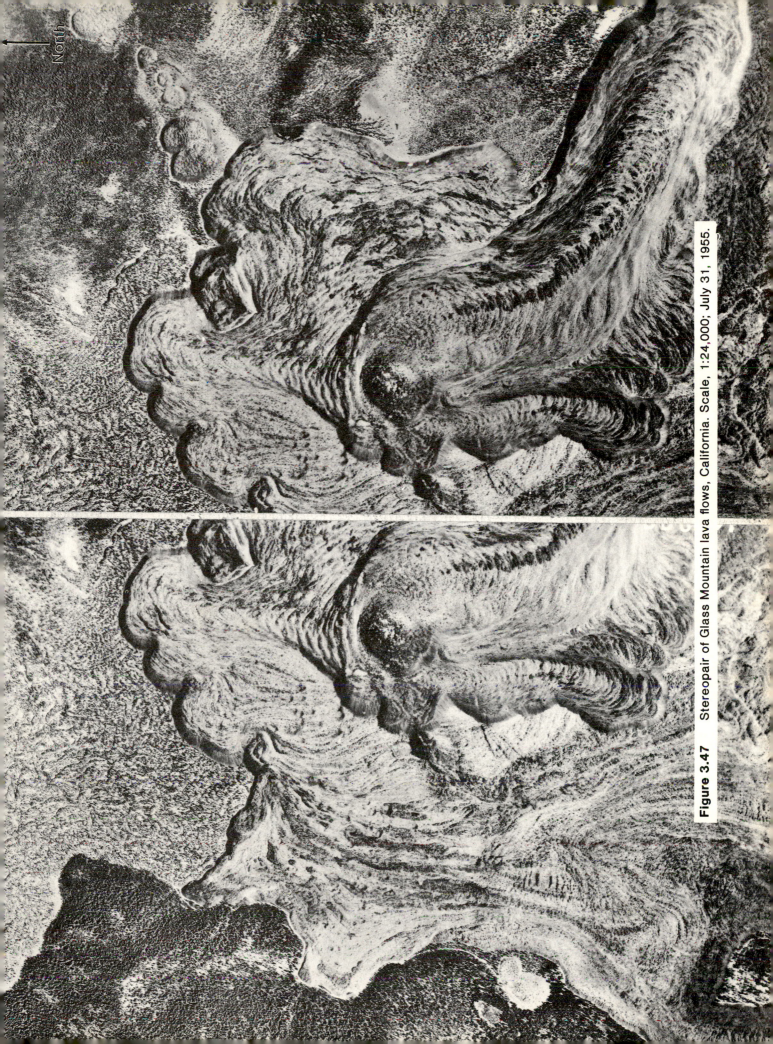

North

Figure 3.47 Stereopair of Glass Mountain lava flows, California. Scale, 1:24,000; July 31, 1955.

19B. PAXTON SPRINGS MAP, New Mexico

A small cinder cone occurs in the center of this map area, and small lava flows are also shown.

1. Study the position of the lava fields with respect to the rest of the terrain. On figure 3.48, trace with a red pencil the borders of the areas covered by lava, using a solid line where the edge of the lava field is not in doubt, and a dashed line where interpolation must be used. What prelava topographic feature guided the flow of the lava as it spread out from its source area? Evidence?

2. Examine the contour lines on and adjacent to the lava field in the area between the town of Paxton Springs and the cinder cone. On the assumption that the present surface of the lava flows has not been eroded appreciably, determine the thickness of the flow and explain how you accomplished it.

3. Study the relationship of the volcanic cone and the lava flows. Write a short statement describing the probable sequence of volcanic events that resulted in the present topographic relationship between the lava flows and the cone.

Figure 3.48 PAXTON SPRINGS MAP
Part of the U.S.G.S. Paxton Springs
Quadrangle, New Mexico, 1952. Scale,
1:24,000; contour interval, 40 feet.

19C. ISLAND OF HAWAII, Hawaiian Islands

The island of Hawaii (fig. 3.50) is the largest of the Hawaiian Islands in the Pacific Ocean, and is composed of five lava domes, the two highest of which are Mauna Kea and Mauna Loa. The summits of these two are more than 13,000 feet above sea level, but the base of the composite lava mountain called Hawaii lies on the floor of the Pacific Ocean. When this fact is taken into consideration, the island of Hawaii rises more than 30,000 feet above its base.

The five lava domes that make up the island are composed of thousands of lava flows that have erupted from an underlying magma. All of the five domes except Kohala have experienced eruptions in historic times. Most of the historic flows of the nineteenth and twentieth centuries associated with Mauna Loa are shown on figure 3.50. These flows were extruded either from the crater at the crest of Mauna Loa called Mokuaweoweo, or large cracks on the southwest and northeast flanks called the Southwest Rift Zone and Northeast Rift Zone, respectively. Figure 3.51B shows many more flows on the flanks of Mauna Loa than are shown on figure 3.50.

1. Figure 3.49 is a topographic profile from the east coast to the west coast of the island of Hawaii across the summit of Mauna Loa.

   a) What is the approximate vertical exaggeration of the profile?

   b) What type of volcano does the island of Hawaii represent? (See fig. 3.51A.)

2. Some of the lava flows associated with Mauna Loa terminate at the coast. What type of geologic feature is formed when lava flows into the ocean?

3. What topographic evidence reflects the presence of more extensive eruptive activity along the Southwest Rift Zone than along the Northeast Rift Zone?

4. Table 3.3 lists in chronological order the dates of the new lava flows from Mauna Loa that appear on figure 3.50. The 1986 flows that reached the coast and over-ran homes and roads were flows from Kilauea, a smaller but separate volcano on the southeast side of Mauna Loa.

   a) What is the mean (average) number of years between all of the eruptions that occurred between 1843 and 1984?

   b) What are the maximum and minimum number of years between two eruptions during this period?

   c) Write a statement either in defense of or in opposition to the following hypothesis: Mauna Loa has erupted with regular periodicity during the period 1843 and 1984, and future eruptions can be predicted on the basis of this periodicity. (Regular periodicity means that all eruptions were separated by more or less equal intervals of time.)

References

Geological Highway Map of Alaska and Hawaii. AAPG, P.O. Box 979, Tulsa, Oklahoma 74101.

Bullard, Fred M. 1963. *Volcanoes: in History, in Theory, in Eruption.* University of Texas Press, Austin, Texas, pp. 217–28.

Table 3.3   List of Years When Lava Flows Occurred on Mauna Loa between 1843 and 1984

1843	1899
1849	1903
1851	1907
1852	1914
1855	1916
1859	1919
1865	1926
1868	1933
1871	1935
1872	1940
1877	1942
1879	1949
1880	1950
1887	1975
1892	1984
1896	

Source: Lockwood, John P., and Peter W. Lipman (1987), Holocene Eruption History of Mauna Loa Volcano, Chapter 18 in R. W. Decker, T. L. Wright, and P. H. Stauffer, eds., *Volcanism in Hawaii,* U.S.G.S. Prof. Paper 1350.

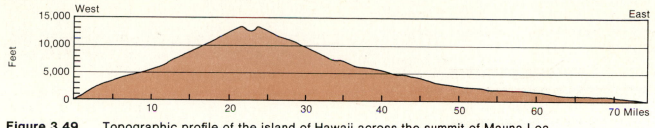

Figure 3.49   Topographic profile of the island of Hawaii across the summit of Mauna Loa.

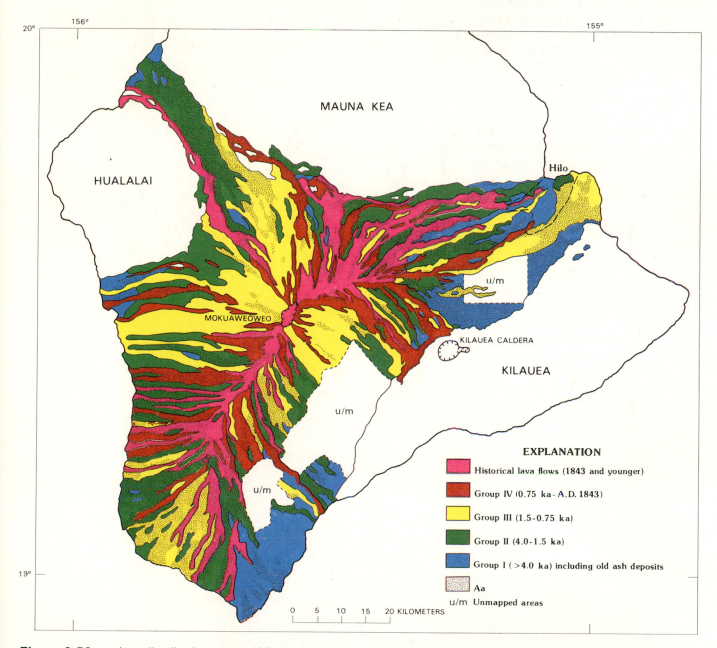

EXPLANATION

- Historical lava flows (1843 and younger)
- Group IV (0.75 ka - A.D. 1843)
- Group III (1.5-0.75 ka)
- Group II (4.0-1.5 ka)
- Group I (>4.0 ka) including old ash deposits
- Aa
- u/m Unmapped areas

Figure 3.50   Age-distribution map of Mauna Loa lava flows, greatly generalized from 1:24,000 mapping, including unpublished work in Hilo area by J. M. Buchanan-Banks. The explanation for this map contains the notation ka, which stands for 1,000 years. Thus, 0.75 ka = 750 years, 1.5 ka = 1,500 years, and 4.0 ka = 4,000 years. (From Lockwood, John P. and Peter W. Lipman (1987), Holocene Eruption History of Mauna Loa Volcano, Chapter 18 in R. W. Decker, T. L. Wright, and P. H. Stauffer, editors, *Volcanism in Hawaii,* U.S.G.S. Prof. Paper 1350.)

Interpretation of Maps, Photographs, and Satellite Images

A

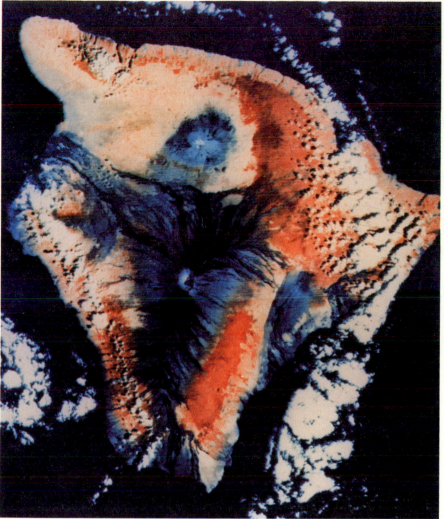

B

Figure 3.51 Mauna Loa, an example of a shield volcano. (*A*) Viewed from low altitudes. Note flat shape, and fresh lava flows visible at left. (*B*) Bird's-eye view of the island of Hawaii, taken by Landsat satellite, shows its volcanic character more clearly. Large peak is Mauna Loa; smaller one, Mauna Kea. Compare with figure 3.50. (*A*) Photo by G. A. MacDonald, U.S.G.S. (*B*) © NASA.

Eruption of Mt. St. Helens

On May 18, 1980, Mt. St. Helens, one of the several spectacular volcanic peaks in the Cascade Mountains of the Pacific Northwest underwent a violent volcanic eruption after two months of low level activity characterized by earthquakes, steam venting, and small ash eruptions. This eruption of Mt. St. Helens, ending a 123-year dormant period, began with seismic activity on March 20, 1980. Steam vents opened up and ash began to be erupted by March 27. Small craters close to the summit developed, ash plumes were erupted, and ash avalanches took place in early April. Seismic activity continued, and by mid-April a significant crater had formed. The summit area of Mt. St. Helens began to swell, enough so that Goat Rocks on the northern flank had a measured movement of 20 feet vertically and 9 feet horizontally to the northwest.

As this activity continued, ash and steam continued to be produced, but in late April seismic activity was greatly reduced. In early May as swelling of the peak continued, the U.S. Geological Survey reported that the northern rim of the crater was rising at a rate of 2 to 4 feet per day.

Through the use of remote sensing techniques utilizing infrared film, hot spots were recorded in early May. Swelling continued, seismic activity increased, and at 8:32 A.M. on May 18, 1980, the violent eruption of Mt. St. Helens occurred, an event that claimed over 60 lives, devastated over 200 square miles of timberland and recreational areas, and spread measurable thickness of ash over several states, and eventually around the world.

The eruption itself was marked initially by an earthquake that triggered a major landslide down the north side of the mountain, followed quickly by a violent explosion. The two events combined to destroy the north rim of the summit crater (Location 7, fig. 3.52). The rock material moved down the north side of the mountain as a mixture of rock, ash, steam, and glacial ice (Location 6, fig. 3.52). Additional fluidization occurred when this avalanche mass hit the water of Spirit Lake and Toutle River. The velocity of the avalanche has been estimated at over 150 mph.

A portion of this enormous avalanche flowed down the valley of the Toutle River for 13 miles, depositing materials in a swath up to 1.2 miles wide and with a thickness up to 450 feet (fig. 3.52). Another part of the avalanche continued to the north, rose over a ridge that was 1,000 feet high, depositing over 100 feet of debris on top of the

ridge before pouring over into the valley of South Coldwater Creek on the north side of the ridge (Location 4, fig. 3.52). New lakes were formed as stream valleys were dammed by debris (Location 3, fig. 3.52), and new islands were formed in Spirit Lake (Location 5, fig. 3.52). The heavy black line on figure 3.52 marks the southern edge of the "Eruption Impact Area" as defined by the U.S. Geological Survey.

Large areas were covered by mudflows (unstippled grey areas, fig. 3.52) that resulted from the mixture of water from melting glaciers and large quantities of ash that poured out during the eruption. The major river draining the area to the north of Mt. St. Helens, the Toutle River, carried large quantities of sediment almost as mudflow to the west into the Cowlitz River and eventually into the Columbia River. Previously recorded flood stages on the Toutle River were exceeded by almost 30 feet. Silting occurred in the Columbia River at the mouth of the Cowlitz River trapping ships upstream. Dredging of a channel was necessary before these ships could move down river to the ocean.

Ash falls occurred to the east of Mt. St. Helens affecting cities such as Yakima, in the heart of the apple growing district of Washington, and the major wheat growing areas farther east. At Ritzville, 205 miles east from Mt. St. Helens, a fine ash deposit of 70 mm was recorded, and by May 21 ash had spread across the continent to the east coast. A later ash eruption on May 25 spread ash to the northwest mantling an area lying roughly between the Columbia River and Olympia, Washington. Small ash eruptions have occurred since, but none as large as the May 25 activity.

The eruption of Mt. St. Helens was the first such volcanic event in the contiguous 48 states since the eruption of Mt. Lassen in Northern California that began in 1914 and continued until 1921. Volcanic activity in the Cascade Range was recorded during the 1800s, and several peaks such as Mt. Ranier, Mt. Baker, and Mt. Hood still have active fumeroles.

Volcanic activity has continued on Mt. St. Helens at a much reduced rate in the years following the 1980 eruption. This activity has included gas and ash emissions, earthquakes, rockfalls, extrusion of a lava spine in the crater, and there is now forming a "composite dome" in the center of the crater. This activity is continually monitored by the U.S. Geological Survey and is providing considerable information that may help in the development of earthquake and volcanic eruption prediction models.

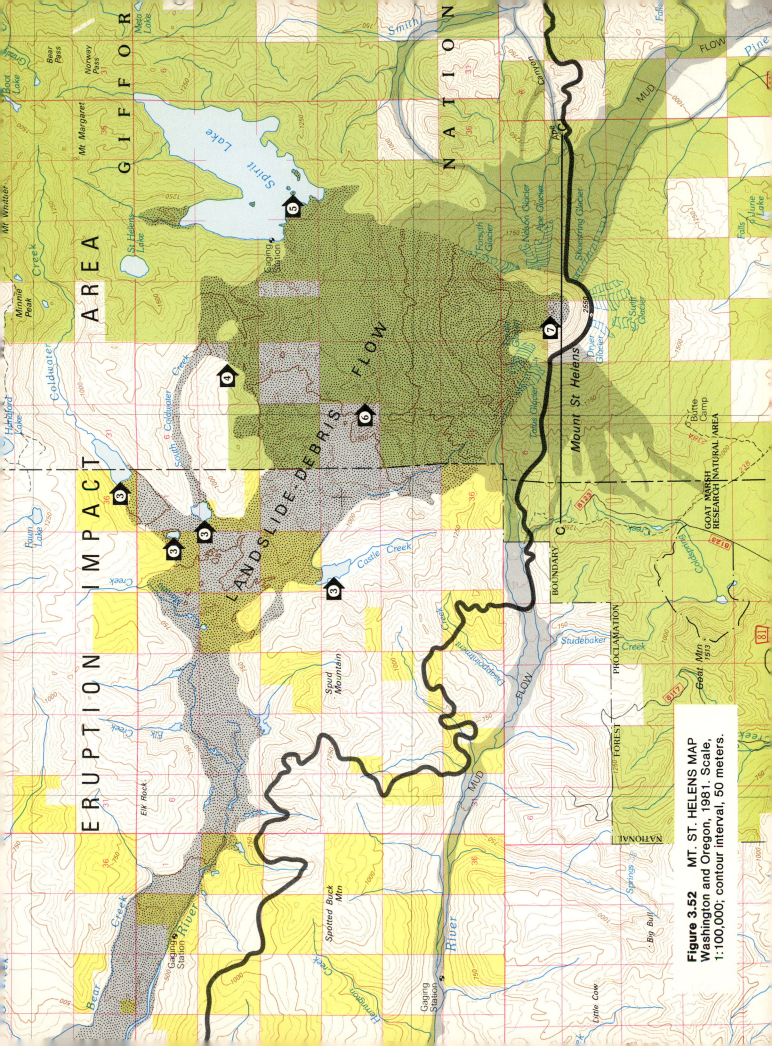

Figure 3.52 MT. ST. HELENS MAP
Washington and Oregon, 1981. Scale,
1:100,000; contour interval, 50 meters.

19D. MT. ST. HELENS, Washington

1. Figure 3.53 is a photograph of the north face of Mt. St. Helens following the eruption of May, 1980. Figures 3.54A and 3.54B are topographic profiles through the summit of Mt. St. Helens showing the pre- and post-eruption profiles of the mountain.

   a) Examine figure 3.53 and the topographic profiles in figure 3.54. Compare the pre-eruption profile of Mt. St. Helens with that of Mauna Loa shown in figure 3.49 (notice the difference in vertical exaggeration between the profiles). How does this volcanic mountain differ from Mauna Loa? What type of volcano is Mt. St. Helens? Mauna Loa?

2. Figure 3.55A is a high altitude infrared photo of Mt. St. Helens and vicinity taken before the May 1980 eruption. Figure 3.55B is a similar photo taken after the eruption. Using the false color photos and the topographic map in figure 3.52 describe the impact of the May 1980 eruption on such elements of the local area as hydrology (lakes, streams, glaciers), logging activities, agriculture, tourism, and recreation.

Figure 3.53   View toward the south of Mt. St. Helens following the eruption of May 18, 1980. (Photo by Austin Post, U.S.G.S.)

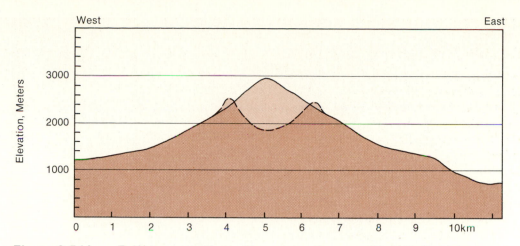

Figure 3.54A   E–W topographic profiles across the summit of Mt. St. Helens showing the pre-eruption profile (solid line), and post-eruption profile (dashed line).

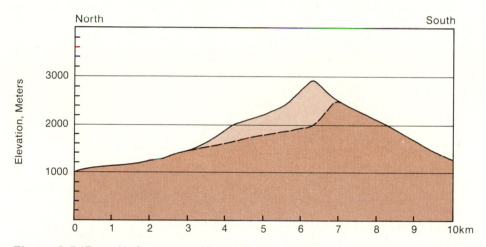

Figure 3.54B   N–S topographic profiles across the summit of Mt. St. Helens showing the pre-eruption profile (solid line), and post-eruption profile (dashed line).

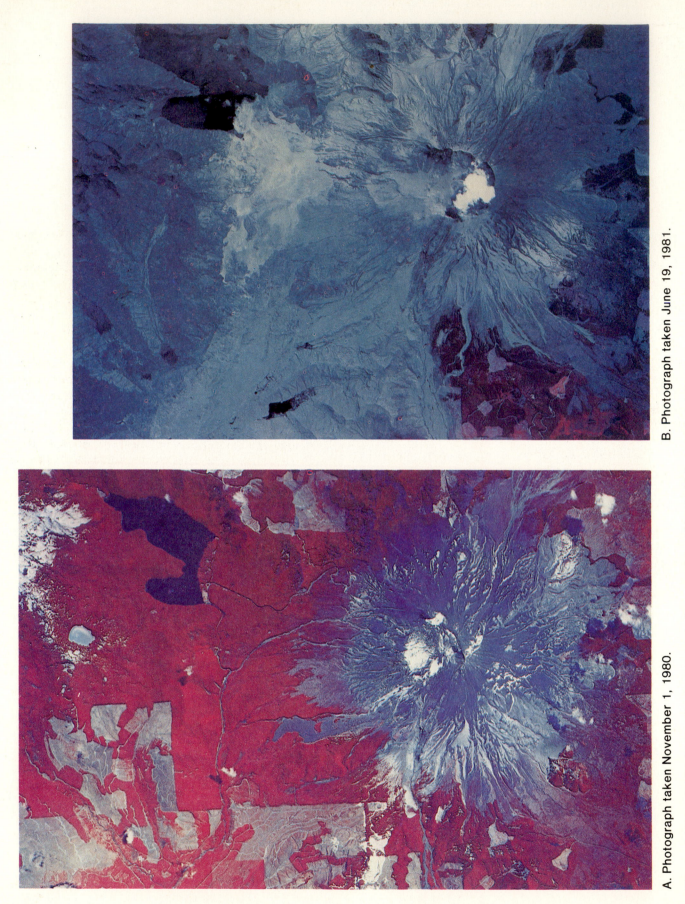

A. Photograph taken November 1, 1980.

B. Photograph taken June 19, 1981.

Figure 3.55   NASA high altitude infra-red photographs of Mt. St. Helens and vicinity.

Structural Geology

Part 4

Introduction

Structural geology deals with the architectural patterns of rock masses as they occur in nature. In Part 1 you were introduced to some basic concepts and principles about the occurrence of rocks. In Part 4, we will build on the concepts and related terminology introduced in Part 1. It may be useful for you to review the text and diagrams on pages 36–40.

Structural geology involves all three rock types—igneous, sedimentary, and metamorphic—but in this part of the manual we concentrate on structures in which sedimentary rock layers are dominant. A sedimentary rock unit that is characterized by a distinct lithologic composition is called a *formation*. A formation is the basic stratigraphic unit depicted on a geologic map. The boundary between two contiguous formations is called a *contact*.

For the sake of simplicity, the formations illustrated herein will be homogeneous in their lithology and will be referred to by some name such as "limestone formation," "sandstone formation," or some other name. In reality, however, a formation may consist of several thinner layers or beds. The contacts between these beds are called *bedding planes* and are more or less parallel to the contacts of the formation itself.

It is common geologic practice to assign a name to a formation. In some cases, the name of the formation will include a lithologic descriptor such as the Madison Limestone or the Pierre Shale. In other cases, the name will consist only of a proper name such as the Nelson Formation or the Sunflower Formation. Formational names such as these will be found only on published geologic maps. Where simple diagrammatic maps are used herein, it will suffice to use only a lithologic definition such as the shale formation or sandstone formation.

Deformation of Sedimentary Strata

During the course of geologic history, sedimentary strata have been subjected to vertical and horizontal forces that may alter the original horizontal position of the rock layers. Some strata may be uplifted in a vertical direction only, so that their original horizontality remains more or less intact. In other cases, the forces of deformation will produce architectural patterns ranging from simple to extremely complex structures.

In order to decipher these structures, geologists measure certain features of a given formation where it crops out at the surface of the earth. These measurements define the position of the formation with respect to a horizontal plane of reference at that particular place. The precise orientation of a contact, bedding plane, or any planar feature associated with a rock mass is called the *attitude*. When attitudes from many outcrops are plotted on a base map such as a topographic map or aerial photograph, and combined with the contacts between formations, the overall geometric pattern or structural configuration of the strata can be determined.

Components of Attitude

The attitude of a formation consists of three parts that collectively define its position at a given location with respect to a horizontal plane and a compass direction.

1. *Strike:* A horizontal line in the plane of the bedding expressed as a compass direction.
2. *Direction of Dip:* The compass direction in which the layer is inclined downward from the horizontal. The direction of dip is always at right angles to the direction of the strike.
3. *Angle of Dip:* The angle between a horizontal plane and a bedding plane. The dip angle is measured in degrees.

The three components of attitude are shown in figure 4.1. As an example of a verbal description of the attitude of a formation at a particular site, the following notation would be used: At the north end of the bridge across the Snake River, the Sundance Formation strikes north 45 degrees east and dips to the southeast at an angle of 30 degrees. On a geologic map, however, the attitude of a formation would be shown by a *strike and dip symbol*. Various forms of this symbol are given in figure 4.2. In illustrations used in parts of this manual, the strike and dip symbols may appear without the notation of the angle of dip.

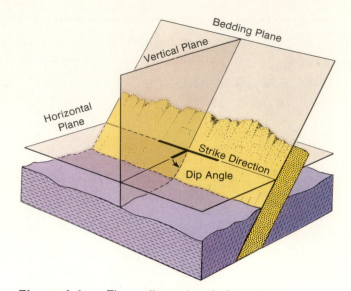

Figure 4.1 Three-dimensional view of an outcropping of sandstone in which the attitude of a bedding plane is measured with respect to horizontal and vertical planes. The shaded slanting plane represents the bedding plane of the layered sandstone. The intersection of the bedding plane and a horizontal plane results in a line called the *strike* of the formation. On a map this line is expressed as a compass direction. The angle formed by the horizontal plane and the bedding plane is the *dip* of the formation. The dip angle is always measured in a vertical plane that is perpendicular to the direction of strike. (From J. H. Zumberge and C. A. Nelson, 1976. *Elements of Physical Geology.* New York: John Wiley and Sons, Inc.)

Methods of Geologic Illustration

Geologic information gathered by the study of outcrops is displayed in a number of ways in order to depict the overall structural features and relative age relations of the strata involved. The three main types of geologic illustrations or diagrams are as follows.

1. *Geologic Map:* This is a map that shows the distribution of geologic formations. Contacts between formations appear as lines and the formations themselves are differentiated by various colors. The map may also show topography by standard contour lines.
2. *Geologic Cross-section:* A diagram in which the geologic formations and other pertinent geologic information are shown in a vertical section. It may also show a topographic profile or it may be schematic and show a flat ground surface.
3. *Block Diagram:* A perspective drawing in which the information on a geologic map and geologic cross-section are combined. This mode of geologic illustration is used to show the three-dimensional aspects of a geologic structure.

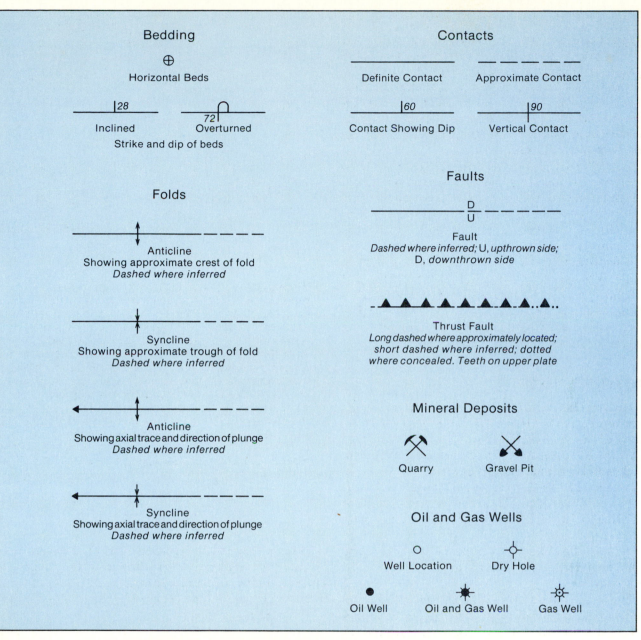

Figure 4.2 Standard symbols used on geologic maps.

Figure 4.3 shows how these three methods of geologic illustrations are related.

Sedimentary Rock Structures

Sedimentary strata that have been subjected to forces of deformation result in three basic geologic structures as shown in the block diagrams of figure 4.4.

1. *Monocline:* A structure in which the strata have a uniform direction of strike but a variable angle of dip.
2. *Anticline:* A structure in the form of an arch.
3. *Syncline:* A structure in the form of a trough.

Notice that in figure 4.4 the arch of the anticline is not reflected in a corresponding topographic arch, and that the synclinal trough is a geologic trough, not a topo-

graphic one. The surface topography of the parallel ridges in figure 4.4*B* and 4.4*C* is controlled by a formation that is more resistant to erosion than the other formations in the structure.

Geometry of Folds

The geometry of a fold is more precisely defined by the attitude of the *axial plane* of the fold, an imaginary plane that separates the *limbs* of the folds into two parts as shown in figure 4.5*A*. The *axial trace* of the fold appears as a line on a geologic map.

If the axial plane is essentially vertical, the fold is said to be *symmetrical* (fig. 4.6*A*); if the axial plane is inclined so that the limbs dip in opposite directions but one limb

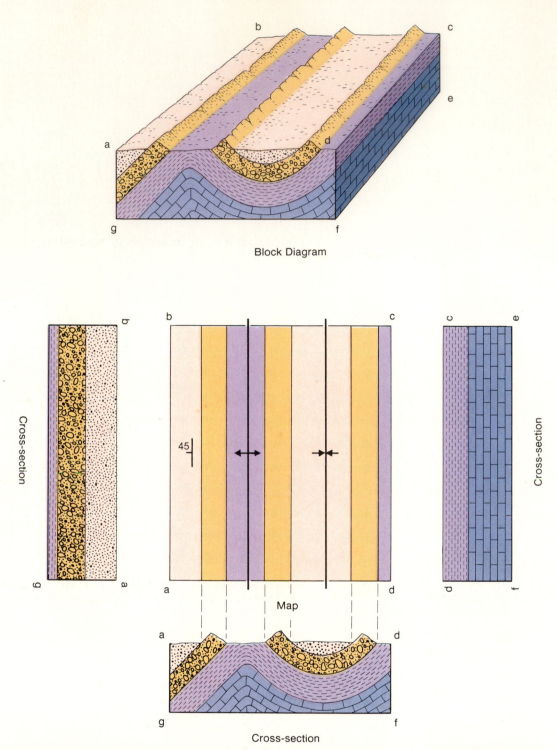

Block Diagram

Cross-section

Cross-section

45

Map

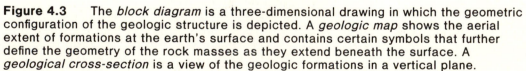

Cross-section

Figure 4.3   The *block diagram* is a three-dimensional drawing in which the geometric configuration of the geologic structure is depicted. A *geologic map* shows the aerial extent of formations at the earth's surface and contains certain symbols that further define the geometry of the rock masses as they extend beneath the surface. A *geological cross-section* is a view of the geologic formations in a vertical plane.

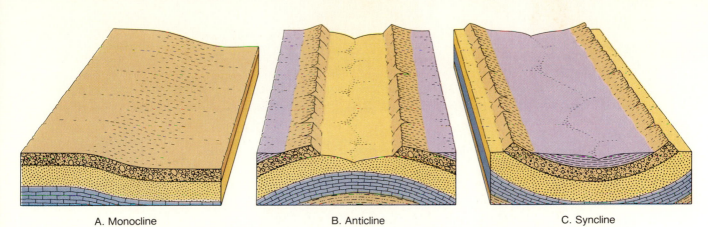

A. Monocline
B. Anticline
C. Syncline

Figure 4.4 Block diagrams of three common folds. The dissected ridges formed by resistant layers in diagrams *B* and *C* are called hogback ridges.

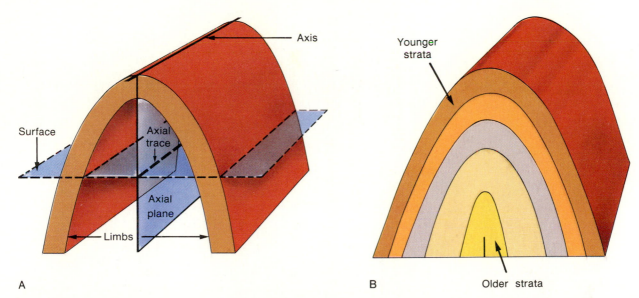

Axis

Surface

Axial trace

Axial plane

Limbs

A

Younger strata

Older strata

B

Figure 4.5 (*A*) Nomenclature of a fold. (*B*) Age relationships of strata in an anticline. (From Carla W. Montgomery, *Physical Geology.* Copyright © 1987 Wm. C. Brown Publishers, Dubuque, Iowa. All Rights Reserved. Reprinted by permission.)

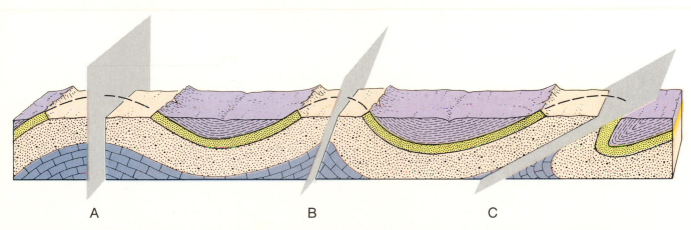

A
B
C

Figure 4.6 Block diagram in which three variations of a fold are shown: (*A*) symmetrical anticline; (*B*) asymmetrical anticline; (*C*) overturned anticline. Note the different attitudes of the three axial planes.

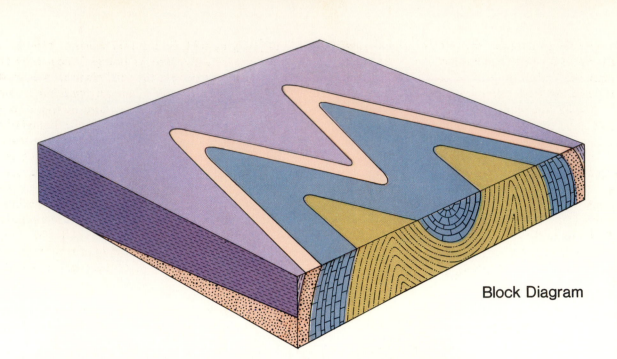

Block Diagram

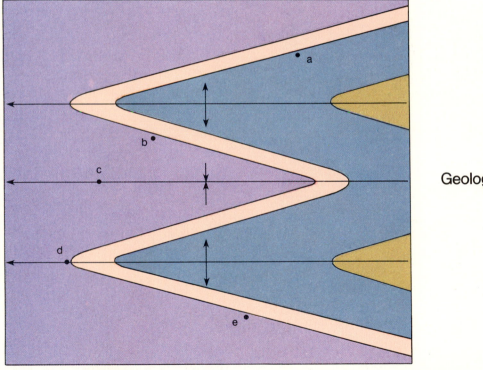

Geologic Map

Figure 4.7   Block diagram and geologic map of plunging folds. The map shows the characteristic outcrop pattern of plunging folds. Here, two plunging anticlines and one syncline plunge to the west (toward the left side of the map). If this area were in a humid region, the arkose formation would be more resistant to erosion than the shale, limestone, or siltstone formations, and would therefore form a hogback ridge.

is steeper than the other, the fold is *asymmetrical* (fig. 4.6*B*); and if the axial plane is inclined to the extent that the opposite limbs dip in the same direction, the fold is *overturned* (fig. 4.6*C*). A *recumbent fold* is an overturned fold in which the axial plane is nearly horizontal. The symbols used on geologic maps to show the traces of axial planes are shown in figure 4.2.

The folds shown in figures 4.4, 4.5, and 4.6 are *nonplunging folds* because the strikes of the limbs are parallel. Another way of describing a nonplunging fold is to say that the strikes of the folded formations are all parallel as shown in figure 4.3.

If, however, the strikes of the formations on either side of an axial plane converge, as in figure 4.7, the fold is said to be a *plunging fold*. A geologic map on which a series

of plunging folds is displayed shows a *"zig-zag"* outcrop pattern. The *direction of plunge* is shown by an arrow placed on the trace of the axial plane as in figure 4.7. *The direction of plunge of a plunging anticline is toward the apex of the converging formations as seen on a geologic map,* and *the direction of plunge of a plunging syncline is toward the open end of the V-shaped pattern of diverging formations.* Figure 4.7 shows both cases.

An anticline that plunges in opposite directions is a *doubly-plunging anticline,* and a syncline that plunges in opposite directions is a *doubly-plunging syncline.* Variations of doubly-plunging folds are the *structural dome* and *structural basin* as shown in figure 4.8. The outcrop patterns of these two structures produce an outcrop pattern of more or less concentric circles (fig. 4.8, *A* and *B*).

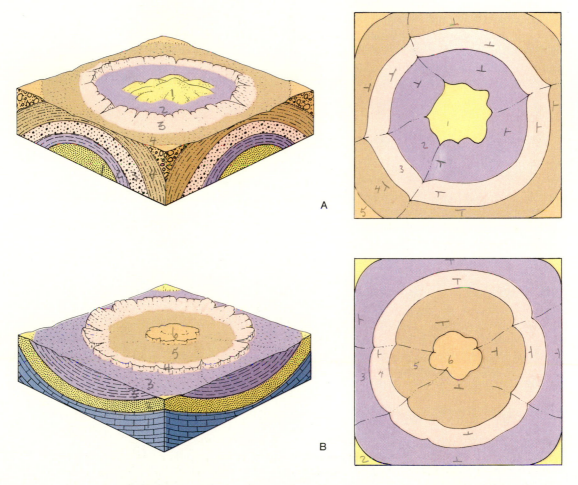

Block Diagram   Geologic Map

Figure 4.8   Block diagrams and geologic maps of two structures that produce circular outcrop patterns: (*A*) structural dome; (*B*) structural basin.

Geologic Maps and Cross-sections

Geologic Maps

A geologic map shows the distribution of rock types in an area. The map is constructed by plotting strikes and dips of formations and the contacts between formations on a base map or aerial photograph. This information is based on field observations on outcrops in the map area. Because a single isolated outcrop rarely yields sufficient information from which the overall structural pattern for a given area can be understood, the geologist must visit enough outcrops in the map area to permit the filling of the gaps from one outcrop to another.

We are not concerned here with the making of actual geologic maps, but rather with their interpretation and the construction of geologic cross-sections from them. The interpretation of a geologic map requires an understanding of the information shown on them and the ability to translate that information onto a geologic cross-section. To accomplish this, one must learn to visualize three-dimensional relationships from a two-dimensional pattern of geologic formations as they appear on a geologic map. This is perhaps the most difficult aspect of structural geology for a beginning student to master, but by following step-by-step instructions, these relationships eventually will become clear.

In general, keep in mind that a geologic map shows the distribution of formations as they appear at the surface of the earth. How this surface information can be used to visualize the unseen components of the rocks below the surface constitutes the subject matter to follow.

Having already been exposed to the notations and symbols used on a geologic map, you may find it useful to review the relationship between a geologic map and a geologic cross-section as shown in figure 4.3. If you thoroughly understand how the geologic cross-section of figure 4.3 relates to the geologic map, and how both relate to the block diagram, you will be in a good position to proceed with the next steps in map interpretation.

Geologic Cross-sections

The purpose of a geologic cross-section is to display geologic features in a vertical section, perpendicular to the ground surface. A crude but nonetheless accurate analogy is to be found in the ordinary layer cake. When a layer cake is viewed from above, all that can be seen is the frosting; the "structure" of the cake is obscured. However, if the cake is cut vertically and the two halves are separated, the component layers of the cake constitute a cross-section of the cake so that its "structure" will be revealed.

Geologic formations do occur in "layer-cake" structures, but they commonly occur in much more complex structures, and it is through the construction of a geologic cross-section that these complexities are unraveled. Following are some general rules and guidelines for use in constructing a geologic cross-section from a geologic map.

1. A geologic cross-section is constructed on a plane that intersects the ground surface at a right angle. The cross-section is shown on the corresponding geologic map by a line that is equivalent to the line along which the cake was cut in the layer cake analogy. Information on or near the line of the cross-section on the map is transferred to the cross-section as the first step in its construction. Such notations as directions and angles of dip, formational contacts, traces of axial planes, and the like provide the basic elements used to make a geologic cross-section.

2. Sedimentary formations to be drawn on cross-sections in the exercises in this manual are assumed to have a constant thickness. That is to say, they do not thicken or thin with depth or along the strike.

3. Dip angles from strike and dip symbols on the map can be used as a basis for estimating the inclination of strata on a cross-section. If dip angles are not shown, keep the dip angles as small as possible but consistent with the thickness of the strata and structural relationships.

4. The relative ages of sedimentary strata in some of the maps and cross-sections used herein are designated by arabic numerals. For example, if four formations are shown on a map or block diagram, the oldest formation is assigned the number "1," and the youngest, a number "4" (fig. 4.9A).

5. If you are required to draw a geologic cross-section from a geologic map on which no strike and dip symbols are present, the direction of dip can be determined in the following manner.

   a) Where a formation contact crosses a stream on the map it forms a V, the apex of which points in the direction of dip as shown in figure 4.9B. (This rule is not to be confused with the "law of V's" as applied to contour lines when they cross a stream.)

   b) The shape of a V formed by a contact that crosses a stream may be used to estimate the angle of dip of the contact. A broad open V indicates a steep dip, and a narrow V is indicative of a shallow dip angle. Where no V is formed, the formation contact is vertical as shown in figure 4.9C. The foregoing method for determining the direction of dip takes precedence over the method described next.

   c) In a sequence of formations, none of which has been overturned, the oldest beds dip toward the youngest as shown in figure 4.9B.

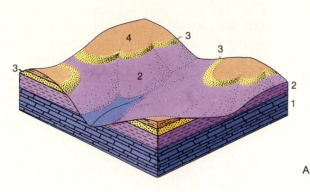
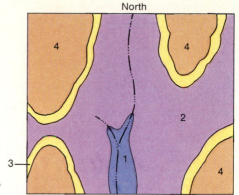

A

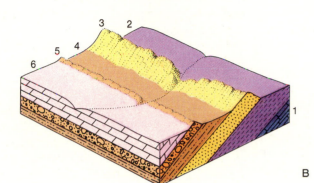
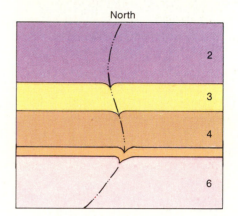

B

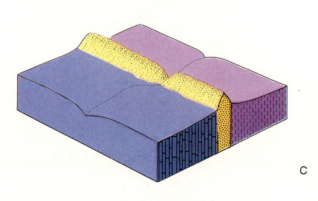
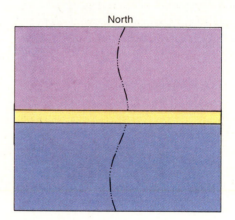

C

Block Diagram

Geologic Map

Figure 4.9 Block diagrams and maps showing relationship of topography to outcrop patterns. (*A*) Horizontal strata dissected by a drainage system. Numbers refer to relative ages of the formations. The formation labeled 1 is the oldest. (*B*) Tilted rock layers showing the V formed where a contact crosses a stream. The oldest beds (i.e., 1, 2, and 3) dip toward the youngest beds (i.e., 5 and 6). The apex of the V's points in the direction of dip. (*C*) Three vertical sedimentary beds, one of which is more resistant to erosion than the other two. In this case the law of V's cannot be used because no V's are formed. The relative age relationship of the three formations cannot be determined from the information shown either on the block diagram or on the map.

Width of Outcrop

When folded strata are exposed to erosion at the earth's surface, they appear as bands on the geologic map. The width of a single band is called *"the width of outcrop"* although the full thickness of the formation may not be exposed in a single outcrop. The width of outcrop is controlled by three factors: the thickness of the formation, the angle of dip of the formation, and the slope of the land surface where the outcrop is exposed.

To illustrate these controlling factors in the simplest case, consider the three horizontal formations of equal thickness in figure 4.10. The geologic cross-section in figure 4.10A shows how the thickness of each formation varies with the slope of the land surface. A gentle slope results in a width of outcrop that is greater than the thickness of the formation, as in the case of the shale formation; and a steeper slope produces a width of outcrop that is less than the thickness of the formation, as in the case of the sandstone and limestone formations.

The lesson to be learned from figure 4.10 is that the width of a formation (width of outcrop) on a geologic map is not necessarily the same as the true thickness of the formation. An exception to this generalization is in the case of a vertical bed, where the width of outcrop is equal to the true thickness of the bed (fig. 4.9C).

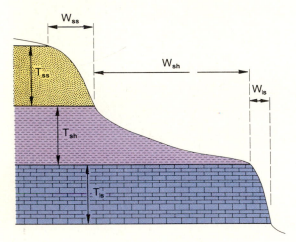

A. Cross-section

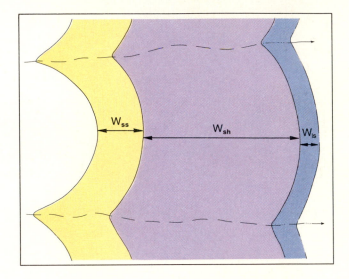

B. Geologic Map

Figure 4.10 (*A*) A cross-section showing three sedimentary formations of equal thickness. T_{ss} = thickness of the sandstone; T_{sh} = thickness of the shale; T_{ls} = thickness of the limestone. The width of outcrop of each formation is greater or less than the corresponding thickness of the formation. ($W_{ss} < T_{ss}$; $W_{sh} > T_{sh}$; $W_{ls} < T_{ls}$.) (*B*) Geologic map showing the width of outcrop of each formation shown on the cross-section in A. The slope of the ground surface is the controlling factor in determining the width of outcrop for each formation.

Exercise 20. Geologic Maps and Cross-sections

1. Complete the four block diagrams in figure 4.11. Below each block diagram, print the name of one or more of the geologic structures shown. Remember that the numbers on the map indicate the relative ages of the formations, with number 1 being the oldest. Assume that the topography in all four diagrams is essentially flat except in diagrams C and D where a stream cuts across the formational contacts.

2. On figure 4.7, place a strike and dip symbol at points *a, b, c, d,* and *e*. Use a black pencil.

3. Figure 4.8 *A* and *B* shows two geologic maps. All formations shown there are sedimentary in origin.
   a) Label each formation with a number indicating its relative age in the sequence of strata. The oldest should be labeled number 1.
   b) Draw strike and dip symbols on each map.

4. On figure 4.10*B*, draw the appropriate geologic symbol that shows the attitude of each of the three formations, and show by numbers the relative age of each.

5. Refer to figure 4.10*A*. On this diagram, none of the three formations has a width of outcrop that is the same as the thickness. In figure 4.9*C* a special case is shown where the width of outcrop of a vertical bed is equal to the thickness of the formation. There is one more special case in which the width of outcrop will equal the thickness of the formation. Using the geologic cross-section of figure 4.10*A* as a base, draw a topographic profile on it in which the width of outcrop of each of the horizontal formations will be the same as the thickness of each. That is, $T_{ss} = W_{ss}$, $T_{sh} = W_{sh}$, and $T_{ls} = W_{ls}$.

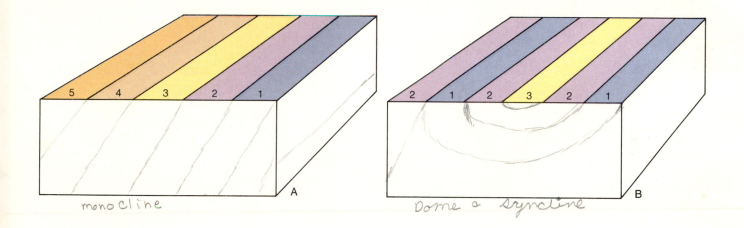

monocline   A

Dome a syncline   B

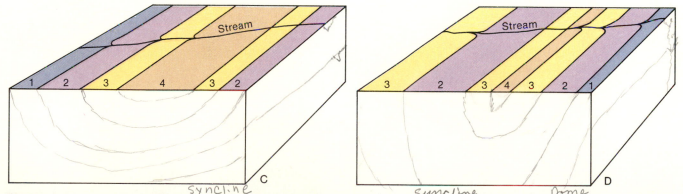

syncline   C

syncline   Dome   D

Figure 4.11   In these four block diagrams, dip directions in *A* and *B* to be determined by the rule that older beds dip toward younger beds, and in *C* and *D*, dip directions are to be determined by the rule of V's.

Exercise 21. Geologic Mapping on Aerial Photographs

21A. CHASE COUNTY, Kansas

Figure 4.12 shows the outcrop pattern of sedimentary strata. The light and dark grey tones correspond to different lithologic characteristics of the various formations. At *A*, near the center of the photograph, the contacts of a light gray formation are shown by two black lines.

1. Extend these lines along the contacts as far as possible on the photo. Do the same for formation *B* shown in the upper left-hand corner of the photo.

2. What is the general attitude of these two formations? (Compare the outcrop pattern of fig. 4.12 with fig. 4.9*A*.)

3. If it is assumed that the thickness of these two formations is constant, why do their widths of outcrop change from place to place?

4. Applying the Law of Superposition in this case, which of the two formations is relatively older than the other?

Figure 4.12   Aerial photograph, Chase County, Kansas. Scale, 1:20,000. (U.S. Dept. of Agriculture photograph.)

21B. FREEMONT COUNTY, Wyoming

This stereopair (fig. 4.13) shows sedimentary strata cropping out in the area. Study the stereopair with a stereoscope, and using figure 4.9B for guidance, answer the following questions.

1. Draw several strike and dip symbols on the right-hand photo of the stereopair, and write a verbal description of the attitude of the formations.

2. Are the oldest beds in the northern or southern part of the area? What rule is applied here that allows you to answer this question?

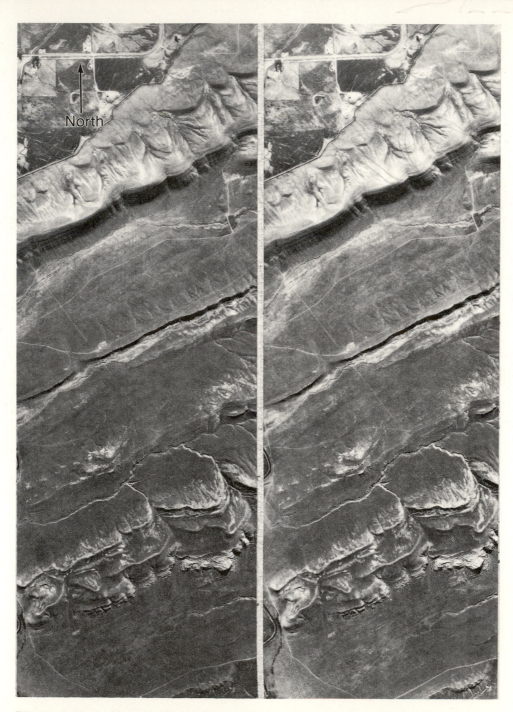

Figure 4.13   Stereopair of part of Freemont County, Wyoming. Scale 1:21,500; July 13, 1960. (U.S.G.S.)

21C. AERIAL PHOTOGRAPH, Arkansas

The pattern of curved ridges in this photograph (fig. 4.14) is the result of differential erosion of sedimentary strata. The ridges are composed of rocks that are more resistant to erosion than the rocks that form the intervening valleys. The structure displayed in the photograph is the nose of a steeply plunging anticline.

1. Draw the trace of the axial plane of the fold on the photograph with a red pencil, and add other appropriate symbols on the axial trace and elsewhere on the photograph to indicate all relevant structural information. (Refer to fig. 4.2 as a reminder of the appropriate symbols to use for the fold axis.)

Figure 4.14   Aerial photograph showing the nose of a plunging anticline in Arkansas. Scale, 1:24,000; November 9, 1957. (U.S.G.S.)

21D. LITTLE DOME, Wyoming

The structure shown here (fig. 4.15) is an elongate dome or a doubly-plunging anticline. Use a stereoscope to study the stereopair while formulating the answers to the following questions.

1. Draw strike and dip symbols on the right-hand photograph of the stereopair. Use red pencil.

2. Draw the trace of the axial plane and other symbols that are appropriate for this structure. Draw on the right-hand photograph and use red pencil.

3. What is the evidence that the angle of dip changes as one follows the ridges northward along the eastern flank of the structure?

4. If a hole were drilled on the axis of the fold at the center of the structure, would the drill encounter any of the formations that crop out on the surface in the area covered by the photographs? Explain how you arrive at your answer.

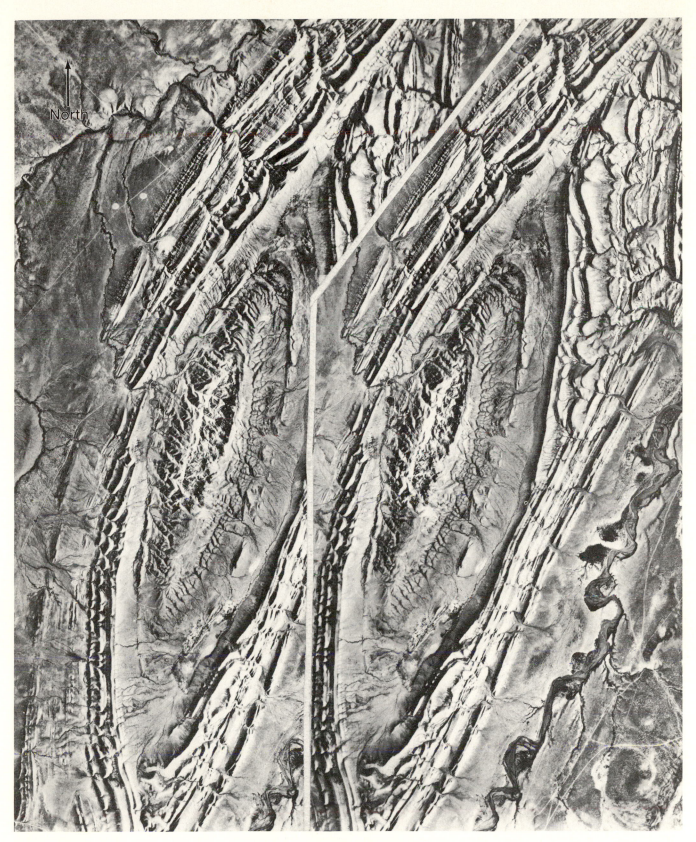

Figure 4.15 Stereopair of Little Dome, Wyoming. Scale, 1:23,600; October 20, 1948. (U.S.G.S.)

21E. HARRISBURG, Pennsylvania

Figure 4.16 looks like an aerial photograph, but in fact is a Side-Looking Airborne Radar (SLAR) image that enhances surface features. A SLAR image is particularly useful in deciphering the structural geology of an area.

The pattern of ridges and valleys shown here are the result of differential erosion of a series of plunging folds in the Appalachian Mountains of Pennsylvania. The river flowing through the area is the Susquehanna River, and the city of Harrisburg lies at the upper margin of the figure near the river.

Figure 4.16 is oriented so that the north direction is toward the lower margin of the figure. The reason for this is that the "shadows" produced by the imaging process must be toward the viewer in order for the ridges on the ground to appear as ridges on the image. (By turning the page upside down, you may see a "reverse topography," that is, the ridges appear as valleys.) Examine the image to get a feel for the structural pattern it reveals.

1. Locate the "zig-zag" ridge that is cut by the Susquehanna River at four places on the image. Using an easily erasable pencil, draw the trend of this formation on the figure. Or, to put it another way, draw a continuous strike line along the crest of the ridge throughout its length. For the purpose of identification, this ridge-forming formation will be called formation A. It forms the flanks of plunging synclines where it is intersected by the Susquehanna River.

2. When you are satisfied that you have identified formation A by the method just described, draw over your pencil line with a yellow felt-tipped pen or pencil to distinguish it clearly from other ridges in the figure.

3. Locate the *next youngest* ridge-forming formation and trace its strike as you did for formation A. This younger formation will be called formation B and it should be identified with a color on the image that contrasts with the one used for formation A.

4. Using a pencil, draw the axial traces of the folds that occur on the figure showing the direction of plunge and the symbol for an anticline or syncline. Reinforce your pencil line with a red pencil when you are certain of your interpretation.

5. Are the rocks west of the Susquehanna River (i.e., between the river and the right-hand margin of the figure) generally older or younger than those near the left-hand margin of the figure? Explain your reasoning.

Figure 4.16 Side-looking Airborne Radar (SLAR) image mosaic of part of the Harrisburg, Pa. Map. (U.S. Geological Survey. Synthetic-Aperture Radar Imagery. Experimental Edition, 1982.) Scale, 1:250,000.

North

Exercise 22. Interpretation of Geologic Maps

22A. LANCASTER GEOLOGIC MAP, Wisconsin

Six formations are shown on this map (fig. 4.17), each of which is identified by an abbreviation. The abbreviations and the names they represent are, in alphabetical order: Od, Decorah Formation; Og, Galena Dolomite; Op, Platteville Formation; Opc, Prairie du Chien Group; Osp, St. Peter Sandstone; and Qal, Alluvium. A *group* consists of two or more formations with significant features in common.

The contacts of these formations are more or less parallel to the topographic contours, thereby indicating that the formations are more or less horizontal. Another set of contours, shown in red, defines the top of the Platteville Formation (Op). The red numbers associated with these red contour lines indicate the elevation of the contour line above sea level.

1. Determine the oldest and youngest formations on the map and those of intermediate age. Complete the geologic column in figure 4.17 by printing the *abbreviation* of a formation in the appropriate box, and printing the *name* of the formation on the line immediately below the box.

2. Using the topographic contour lines, estimate the thickness of the Decorah Formation (Od), the Platteville Formation (Op), and the St. Peter Sandstone (Osp). Print the estimated thickness in feet of each of these formations to the right of the appropriate box in the geologic column. The thickness of a formation is determined by subtracting the elevation of the bottom of the formation from the elevation of the top of the formation. These elevations can be estimated from contour lines on either side of a contact.

3. Why is it impossible to determine the thickness of the Galena Dolomite and the Prairie du Chien Group?

4. Locate Cement School in the NE part of the map area. A road intersection near the school has an elevation of 1,076 feet and is so marked on the map. Using a nearby contour line showing the top of the Platteville Formation, determine how deep a well must be drilled at the road intersection to reach the top of the Platteville Formation.

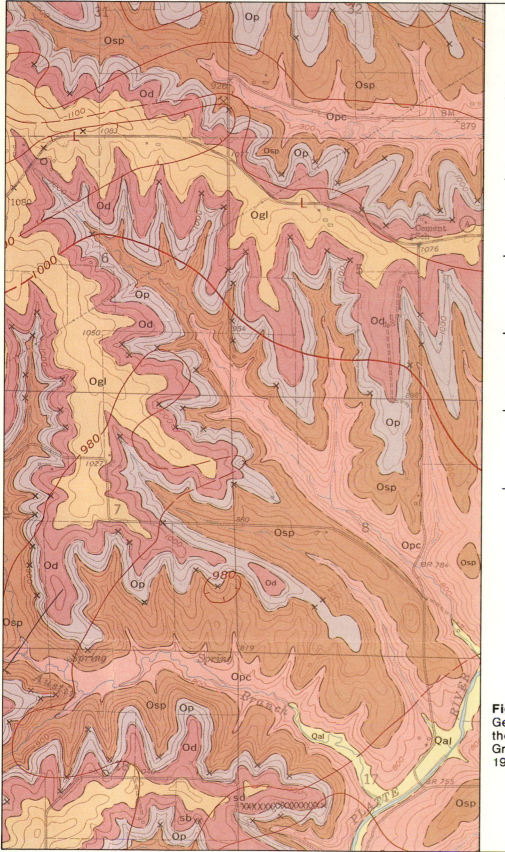

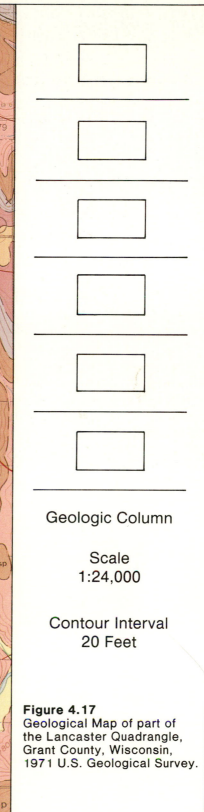

Geologic Column

Scale
1:24,000

Contour Interval
20 Feet

Figure 4.17
Geological Map of part of
the Lancaster Quadrangle,
Grant County, Wisconsin,
1971 U.S. Geological Survey.

22B. SWAN ISLAND GEOLOGIC MAP, Tennessee

This geologic map (fig. 4.19) shows a banded outcrop pattern of sedimentary rocks. Study the map to get a general "feel" for the geology. Note especially the strike and dip symbols on the map.* (North is toward the bound edge of the page.)

1. The topographic profile of figure 4.18 is drawn along the line trending in a northwest-southeast direction across the map from one margin to the other. Draw a geologic cross-section along this line on the profile of figure 4.18. (In aligning the topographic profile with the line of profile on the map, place the point on the profile labeled Briar Fork at the point where Briar Fork crosses the line of profile on the map. Only the word "Fork" appears on the map.) Review the directions on p. 150 before proceeding, and follow them carefully. Regard formations A, B, C, D, EFG, H, I, J, K, and LM as a single formation when you draw the geologic cross-section. Color formations Є̵mn, Ok, C, Os, and MDc with appropriate colors on the cross-section. Ignore the formation labeled Qal. Identify each formation on the cross-section by the appropriate symbol along the topographic profile.

*The number that appears next to a strike and dip symbol on a geologic map refers to the angle of dip of the rock layers as measured by a field geologist at a specific rock outcrop or exposure. When these numbers occur near a line on the map along which a geologic cross-section is to be constructed, they should be considered approximations of the dip angles rather than absolute values. Dip angles within a few tens of feet of each other can vary as much as five to ten degrees.

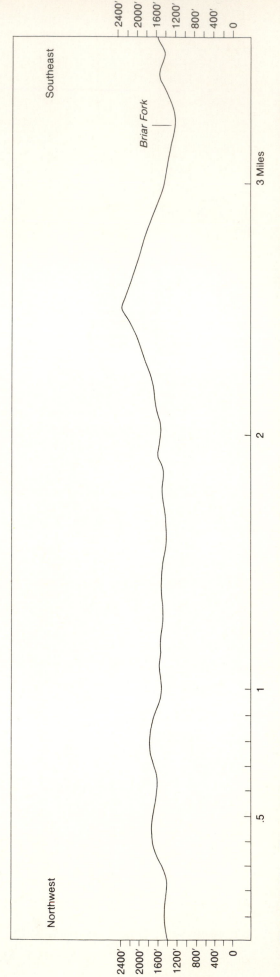

Figure 4.18   Topographic profile from northwest to southeast along the black line shown in the Swan Island Map. (See fig. 4.19 for location.)

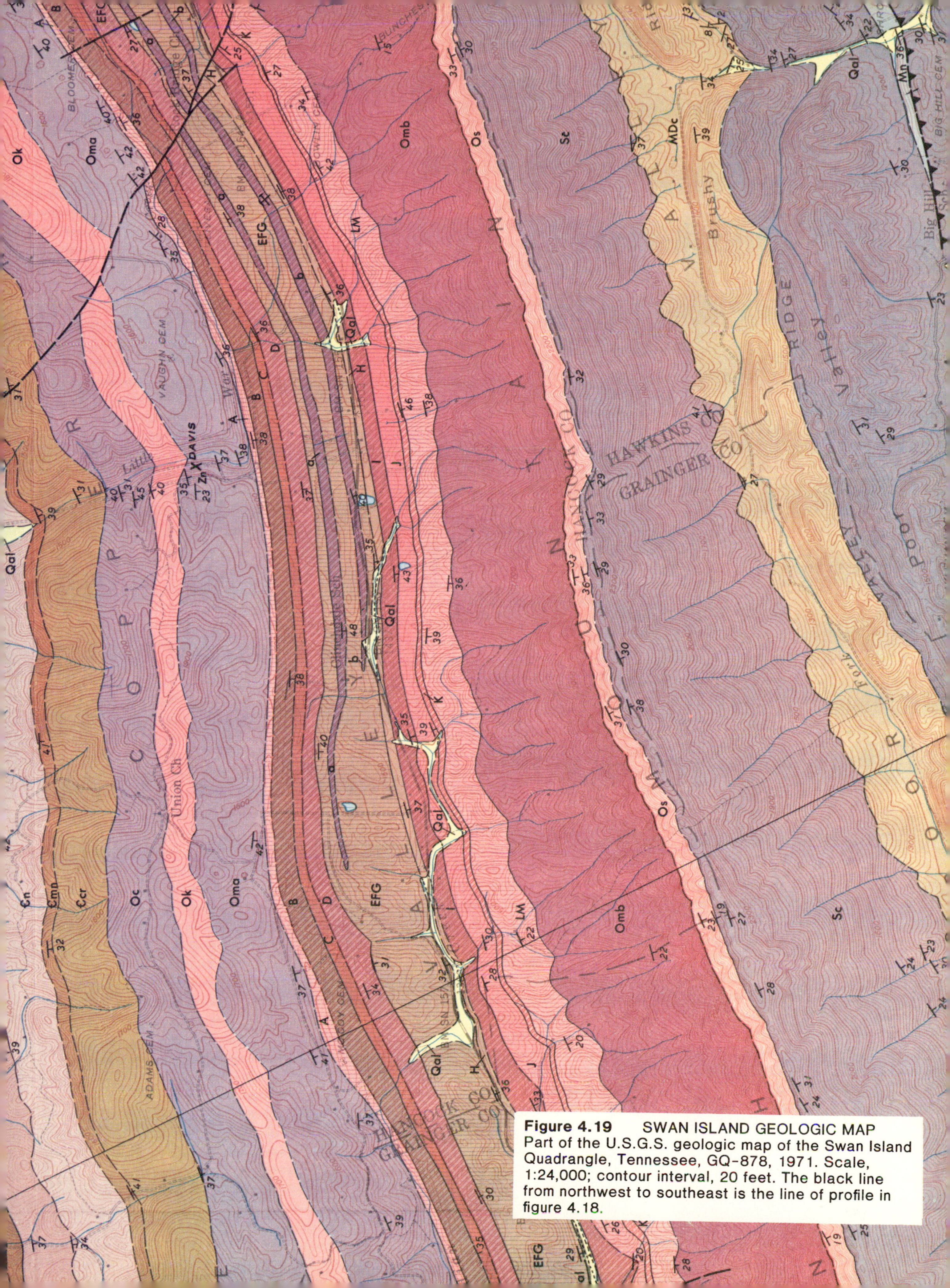

Figure 4.19 SWAN ISLAND GEOLOGIC MAP
Part of the U.S.G.S. geologic map of the Swan Island
Quadrangle, Tennessee, GQ–878, 1971. Scale,
1:24,000; contour interval, 20 feet. The black line
from northwest to southeast is the line of profile in
figure 4.18.

22C. COLEMAN GAP GEOLOGIC MAP,
Tennessee-Virginia

This area (fig. 4.21) is underlain by sedimentary rock formations of different thicknesses. Note the many strike and dip symbols that occur on the map. (North is toward the bound edge of the page.)

1. The topographic profile of figure 4.20 is drawn along the line trending northwest-southeast across the map area from margin to margin, passing through the location of Brooks Well. Draw a geologic cross-section along this line. Align the Brooks Well on the profile with the position of the Brooks Well on the map to achieve the proper correlation between the topography of the profile and the contours on the map. Note that the vertical scale of the profile is identical with the horizontal scale of the map. For assistance in drawing the geologic cross-section, it should be noted that the Brooks Well intersected the contact between the base of the €c formation and the top of the €r formation (not shown on the map) 300 feet below the ground surface. Color formations €c and €cr on the cross-section, and label all formations with their correct symbols.

2. What is the thickness of formation €cr?

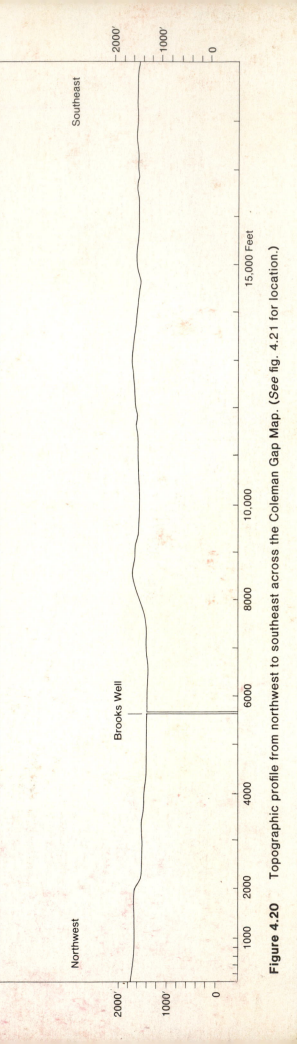

Figure 4.20 Topographic profile from northwest to southeast across the Coleman Gap Map. (*See* fig. 4.21 for location.)

Figure 4.21   COLEMAN GAP MAP
Part of the U.S.G.S. geologic map of the Coleman Gap Quadrangle, Tennessee-Virginia, GQ–188, 1962. Scale, 1:24,000; contour interval, 20 feet. Black line from northwest to southeast is the line of profile in figure 4.20.

Faults and Earthquakes

Earth stresses that produce folds also produce faults. A *fault* is a fracture or break in the earth's crust along which differential movement of the rock masses has occurred. Movement along a fault causes dislocation of the rock masses on either side of the fault so that the contacts between formations are terminated abruptly at the fault.

A fault is a planar feature, and therefore its attitude can be described in much the same way that any geologic planar feature can be described. Figure 4.2 shows the various symbols used on a geologic map to define faults.

Nomenclature of Faults

Figure 4.22 is a block diagram of a hypothetical faulted segment of the earth's crust. The *fault plane* is defined as abcd. The fault plane strikes north-south and dips steeply to the east. A single horizontal sedimentary bed acts as a reference marker and shows that the displacement along the fault plane is equal to the distance x-y. This is called the *net slip*. The arrows show the direction of relative movement along the fault plane. Block A has moved up with respect to block B, and conversely, block B has moved down with respect to block A. Block A is called the *upthrow side* of the fault, and block B is the *downthrow side*. Block B is also known as the *hanging wall*, and block A as the *footwall*. Both terms are derived from miners who drove tunnels along fault planes to mine ore that had been emplaced there.

Faults generally disrupt the continuity or sequence of sedimentary strata and they cause the dislocation of other rock units from their prefaulted positions. On geologic maps, the intersection of the fault plane with the ground surface is called a *fault trace*. Fault traces are depicted on geologic maps by the use of standard symbols (fig. 4.2).

After faulting occurs, erosion usually destroys the surface evidence of the fault plane, so that with the passage of time the *fault scarp* (the exposed surface of the fault plane in figure 4.22) is destroyed. Only the fault trace remains.

Types of Faults

Faults are divided into three major types, each of which is defined by the relative displacement along the fault plane. In the first two types, the main element of displacement has been vertical, more or less parallel to the dip of the fault plane. A *normal fault* is one in which the hanging wall has moved down relative to the footwall (fig. 4.23*A*). A *reverse fault* is one in which the hanging wall has moved up relative to the footwall (fig. 4.23*B*). A reverse fault in which the fault plane dips less than 45 degrees is called a *thrust fault*.

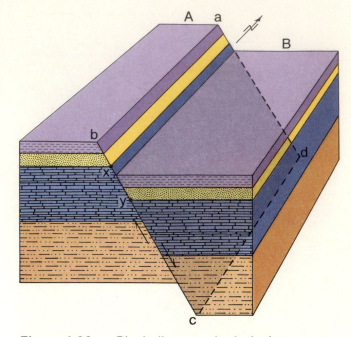

Figure 4.22 Block diagram of a fault. Arrows show the relative movement of block A with respect to block B. The horizontal beds have been dislocated a distance of x-y. The fault plane is abcd.

The third category of faults is characterized by relative displacement along the fault plane in a horizontal direction parallel to the strike of the fault plane. This type of fault is called a *strike-slip fault* (fig. 4.23*C*). Figure 4.23*D* shows a *horst*, an upthrown block bounded on its sides by normal faults. Figure 4.23*E* shows a *graben*, a downthrown block bounded on its sides by normal faults.

A fault shown on a geologic map can be analyzed to determine what kind of fault is involved. The analysis of normal and reverse faults will reveal the hanging and footwalls that lead to an understanding of the relative movement along the fault plane. In a strike-slip fault, offsetting of a marker bed as in figure 4.23*C* is the most direct evidence of the direction of movement. In some cases, the distinction between a strike-slip fault and a normal or reverse fault requires information not shown on the map.

A normal or reverse fault that cuts across the strike of inclined or folded sedimentary beds presents one of the most common situations for the analysis of movement along the fault plane. In such cases, there will be an apparent migration of the beds in the direction of dip of these beds on the upthrow side of the fault as erosion progresses. Stated another way, if an observer were to stand astride the fault trace, the observer's foot resting on the older rock would rest on the upthrow side. This is a simple mental test that can be applied to the analyses of faults presented in Exercise 23.

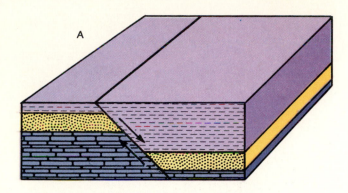

Normal Fault

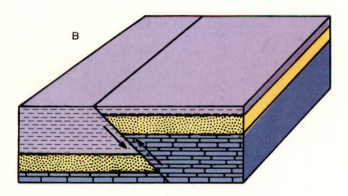

Reverse Fault

Strike-slip Fault

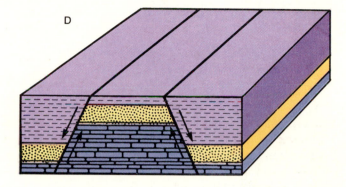

Horst

Graben

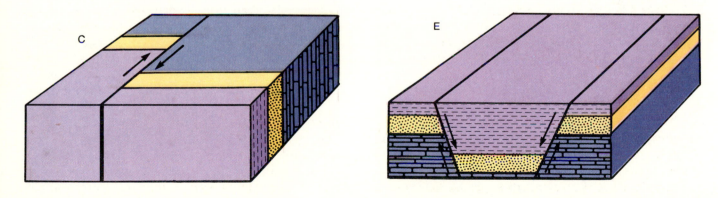

Figure 4.23 Block diagrams illustrating major fault types. Arrows indicate relative movement along the fault plane.

Exercise 23. Fault Problems

23A. FAULT PROBLEMS ON BLOCK DIAGRAMS

1. In figure 4.24, the upper group of three block diagrams shows (A) an unfaulted segment of the earth's crust with an incipient fault plane (dashed line), (B) movement along the fault plane, and (C) the appearance of the faulted area after erosion has reduced it to a relatively flat surface. The sequence D, E, and F is similar. Complete the outcrop patterns of the inclined formation on block diagrams C and F and label each as to the kind of fault (normal or reverse). Indicate by letters the hanging wall (H), the footwall (F), the upthrow (U), and the downthrow side (D) on each side of the fault trace in diagrams C and F.

2. Figure 4.25 shows geologic maps A and B, and their corresponding block diagrams. The sedimentary formations are numbered according to their ages (stratum 1 is the oldest). Complete the block diagram below each map and indicate by letters on the map: the hanging wall (H), footwall (F), upthrow side (U), downthrow side (D), and show by arrows the relative movement along the fault plane. Label each fault as either normal or reverse.

3. Figure 4.26 shows two separate sequences of three block diagrams: A, B, C; and E, F, G. A and E show conditions prior to faulting, B and F show conditions immediately after faulting, and G show the relationships on both sides of the fault traces after the fault scarp has been removed by erosion.

   a) Figure 4.26D is a geologic map of block diagram C. With a black pencil, draw the following symbols on the map: direction of dip of the fault plane; strike and dip symbols on the sedimentary beds on both sides of the fault trace; the upthrow (U) and downthrow (D) sides of the fault; and write either "normal" or "reverse" after the letter D at the lower margin of the map. (Refer to figure 4.2 to refresh your memory as to the various geologic map symbols.)

   b) In the blank space of figure 4.26H, draw a geologic map of block diagram G, draw the corresponding geologic map symbols as in 4.26D, and write either "normal" or "reverse" after the letter H on the lower margin of the map.

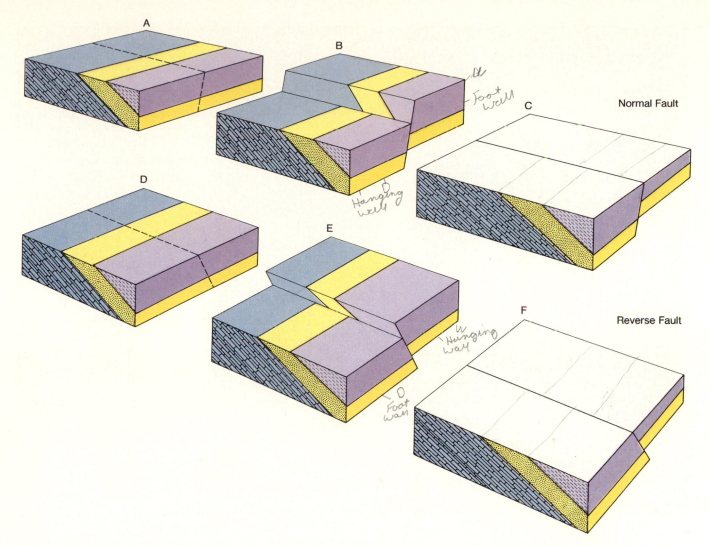

A

B

Foot
Wall

Hanging
wall

Normal Fault

C

D

E

Hanging
way

Foot
wall

Reverse Fault

F

Figure 4.24   Series of block diagrams showing successive stages in the development of both a normal and a reverse fault. A and D before faulting with incipient fault plane marked by dashed line. B and E are shown immediately after faulting. C and F are shown after erosion of upthrow sides to a level common with the downthrown sides.

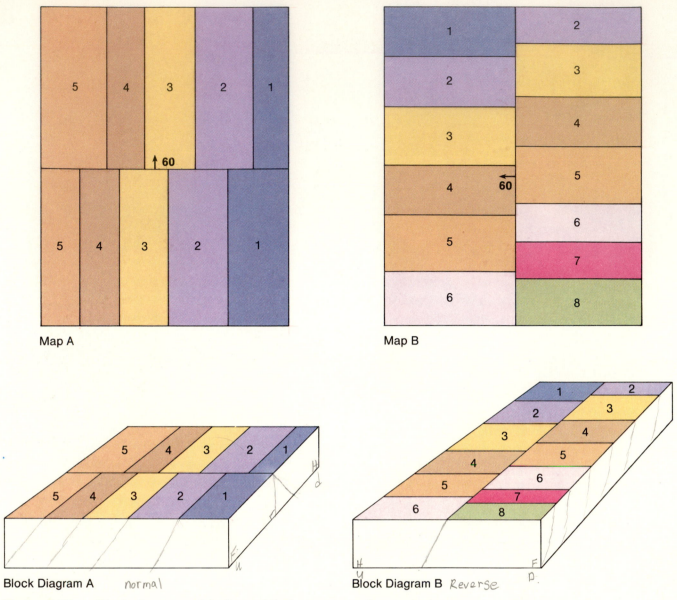

Figure 4.25 Block diagrams of faults and their corresponding geologic maps.

Map A

Map B

Block Diagram A   *normal*

Block Diagram B   *Reverse*

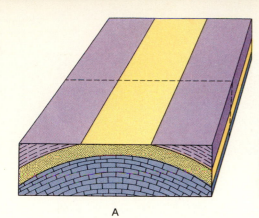

A

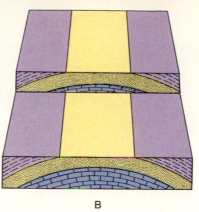

B

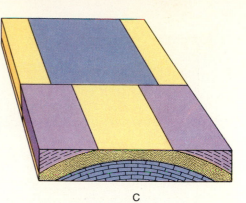

C

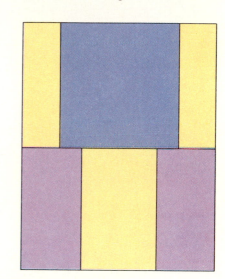

D

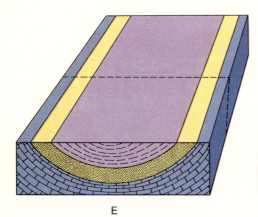

E

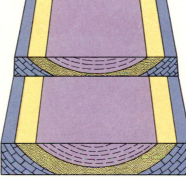

F

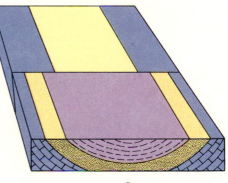

G

H

Figure 4.26 Block diagrams of a faulted anticline and faulted syncline.

23B. SWAN ISLAND GEOLOGIC MAP, Tennessee

1. On the Swan Island Map (fig 4.19), formation Ok is cut by a fault trending NW-SE in the northeast corner of the map area. (North is toward the left margin of the page.) Label the upthrow and downthrow sides of the fault with the correct geologic symbols.

2. Refer to the fault described under question 1. What would be the direction of *dip* of the fault plane if this were a normal fault?

3. If the fault referred to in question 1 were a reverse fault, label the hanging wall side with an H.

23C. LANDSAT FALSE COLOR IMAGE OF THE LOS ANGELES AREA, California

The image of figure 4.27 extends from the Mohave Desert on the north to the Pacific Ocean on the south. Figure 4.28 is a generalized map of the same area showing the traces of *some* of the faults in the area. Those that are easily visible on the image are shown with a solid line, and those that are more difficult to recognize are shown by dashed lines.

1. Transfer the fault traces from figure 4.28 to figure 4.27.

2. Describe the physiographic features associated with the San Andreas and Garlock faults.

3. The San Andreas fault is a strike-slip fault that extends from the Gulf of California to beyond San Francisco in the Pacific Ocean, a distance of about 600 miles. The Pacific Ocean side of the fault has moved north (west in the area of the image) some 300 to 350 miles in a series of horizontal displacements. The San Francisco earthquake of 1906 was caused by slippage along the San Andreas fault in the amount of 21 feet.

   Figure 4.28 shows the epicenters of two major earthquakes in the greater Los Angeles area during historical times, the Ft. Tejon earthquake of 1857 and the Sylmar earthquake of 1971. The *epicenter* of an earthquake is a point on the earth's surface directly above the focus. The *focus* is the place of origin of an earthquake beneath the earth's surface. The focus lies on the fault plane associated with the earthquake. In a cross-section, a line drawn through the focus and the surface trace of the fault that caused the earthquake defines the fault plane.

a) Figure 4.29 is a schematic cross-section of the earth across the San Andreas fault. The focus and the epicenter of the Ft. Tejon earthquake are shown. Draw a solid red line on the diagram showing the attitude of the San Andreas fault. What is the dip of the fault plane?

b) Figure 4.30 is a schematic cross-section through the focus of the 1971 Sylmar earthquake to the fault scarp caused by the Sylmar earthquake. Precise surveying after the Sylmar earthquake showed that the San Gabriel Mountains increased about 6 feet in elevation as a result of the movement along the San Fernando fault. This movement was responsible for the Sylmar earthquake.

   Draw a solid red line on figure 4.30 showing the San Fernando fault plane. Draw red arrows on each side of the fault indicating the relative movement along the fault plane. Label the hanging wall (H) and the footwall (F). Is the San Fernando fault a normal, reverse, or strike-slip fault? How determined?

Reference

Greensfelder, Roger. 1971. Seismologic and crustal movement investigations of the San Fernando earthquake. *California Geology,* April-May 1971: 62–68. California Division of Mines and Geology, Sacramento, California 95814.

Figure 4.27   False color image of the greater Los Angeles area of southern California, made from Landsat 1, October 21, 1972. (NASA ERTS E–1090–18012.)

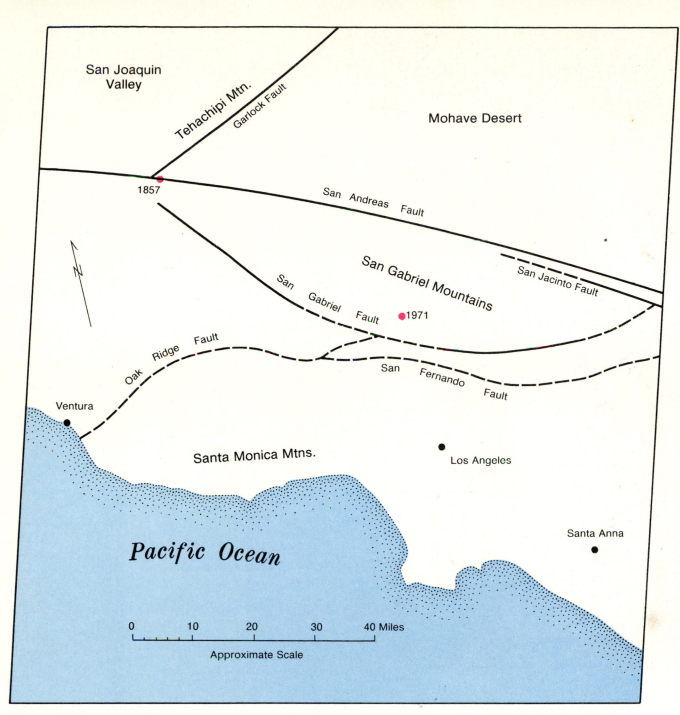

San Joaquin
Valley

Mohave Desert

Tehachipi Mtn.

Garlock Fault

1857

San   Andreas   Fault

San Gabriel Mountains

San Jacinto Fault

San   Gabriel   Fault

●1971

Oak   Ridge   Fault

San   Fernando   Fault

Ventura

Santa Monica Mtns.

Los Angeles

Santa Anna

Pacific Ocean

0   10   20   30   40 Miles

Approximate Scale

Figure 4.28   Generalized map of the greater Los Angeles area showing traces of some of the faults
occurring there, also the epicenters of the Ft. Tejon earthquake of 1857 and the Sylmar earthquake of
1971. (R. H. Campbell. 1976. Active faults in the Los Angeles-Ventura area of southern California.
ERTS-1 A New Window on Our Planet, U.S.G.S. Professional Paper 929, pp. 113-16.)

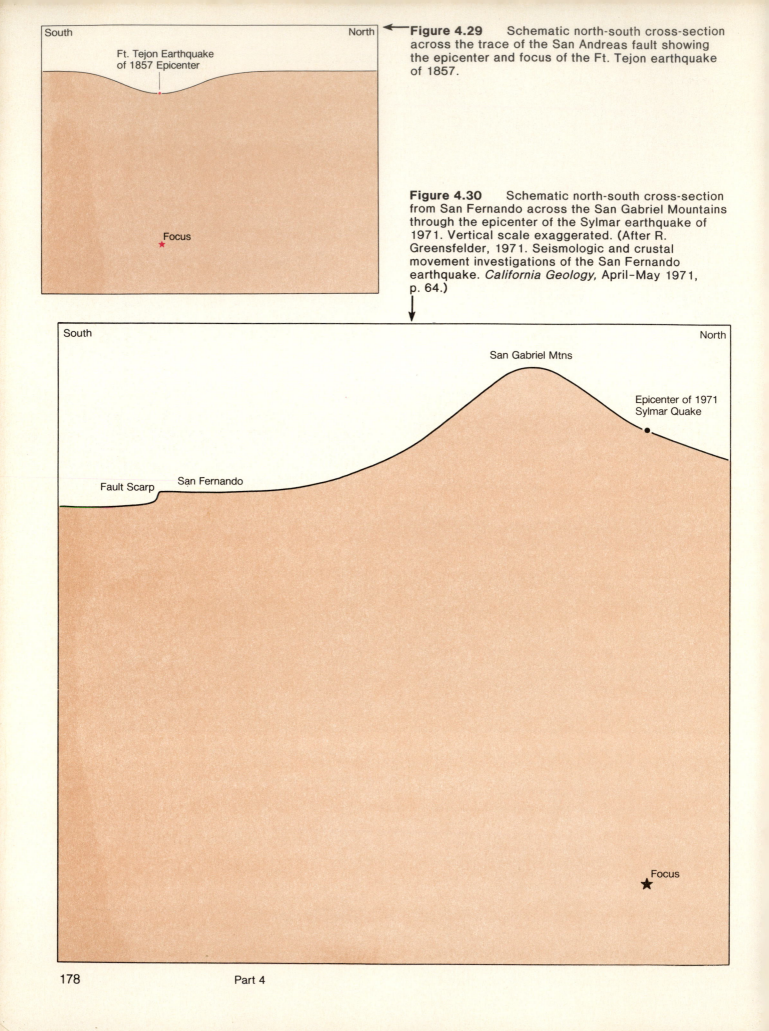

Figure 4.29 Schematic north-south cross-section across the trace of the San Andreas fault showing the epicenter and focus of the Ft. Tejon earthquake of 1857.

Figure 4.30 Schematic north-south cross-section from San Fernando across the San Gabriel Mountains through the epicenter of the Sylmar earthquake of 1971. Vertical scale exaggerated. (After R. Greensfelder, 1971. Seismologic and crustal movement investigations of the San Fernando earthquake. *California Geology,* April–May 1971, p. 64.)

South   North
Ft. Tejon Earthquake of 1857 Epicenter
Focus

South   North
San Gabriel Mtns
Epicenter of 1971 Sylmar Quake
Fault Scarp   San Fernando
Focus

Plate Tectonics and Related Geologic Phenomena

Part 5

Introduction

The theory of plate tectonics has gained widespread acceptance only since the mid-1960s. Before then, most geologists in the Northern Hemisphere believed that the continents and ocean basins had been fixed in more or less their present positions for most of geologic time. The term *plate tectonics* refers to the rigid plates that comprise the skin of the earth and their movement with respect to one another.

Figure 5.1 shows the distribution of the plates as they are currently postulated. The plates differ greatly in size, and some include both continents and ocean basins within their borders, for example, the Africa and North America plates.

Plate tectonics explains many phenomena on the planet Earth, both on the continents and in the ocean basins, such as the origin and distribution of volcanoes and earthquakes, the topography of the sea floor, and a host of other major geologic features that seemed unrelated before the concept was proposed. Our interest in plate tectonics will be directed mainly toward the characteristics of *plate boundaries,* or *margins,* and the processes that occur there.

The dynamics of plate boundaries explains many geologic events and features on a global basis. About a dozen different types of plate boundaries have been identified in the literature, but in the exercises that follow only two will be dealt with. Before proceeding, however, it will be useful to review some basic terminology associated with the earth's outer layers.

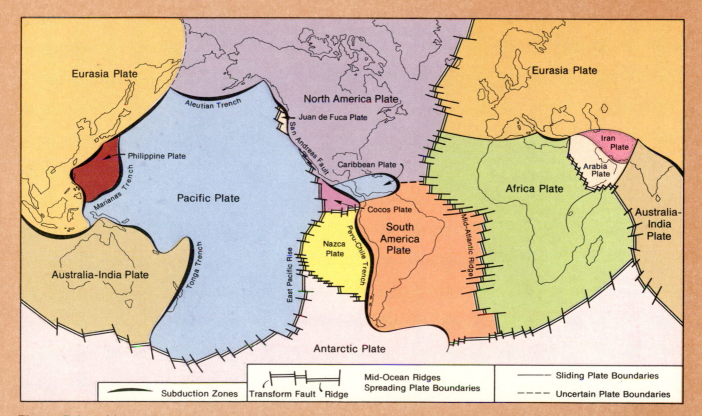

Figure 5.1   Map of the earth showing the names and boundaries of lithospheric plates of the earth's crust. (Compiled from various sources.)

The Lithosphere and the Asthenosphere

The earth's outermost layer is called the *lithosphere.* It consists of an upper part called the *crust,* and a lower part called the *upper mantle.* The *asthenosphere* lies beneath the lithosphere and consists of material that is plastic in its behaviour; it tends to flow under stress rather than fracture like rocks in the crust. The *lower mantle* forms the upper part of the asthenosphere, and it is along the boundary between the upper and lower mantles that plate movement is believed to occur.

Figure 5.2 shows the relationship of crust and mantle to the lithosphere and the asthenosphere as they exist beneath the continents and ocean basins. Notice that *continental crust* is considerably thicker than *oceanic crust,* and that continental crust is essentially granitic in composition, whereas oceanic crust is basaltic. The difference in density between granite and basalt explains why the continents generally lie at a higher elevation than the ocean basins with respect to sea level.

The cause of the movement of lithospheric plates is not well understood, but it is believed that the forces originate in the asthenosphere, and the lithospheric plates are simply rafted along as the asthenosphere shifts around. The most common explanation of plate movement is that differential heating in the asthenosphere produces giant convection currents that drive the process. Our interest here lies with the results of plate movement rather than the causes.

Plate Boundaries

A variety of conditions prevail at different times along plate boundaries. Some plate margins are *passive,* which means that the plates involved are essentially static and not in motion. Other margins are *active* and moving in different directions. Two plates may slide past each other in a shearing motion, abut against each other in sort of a slow motion collision, or pull away from each other.

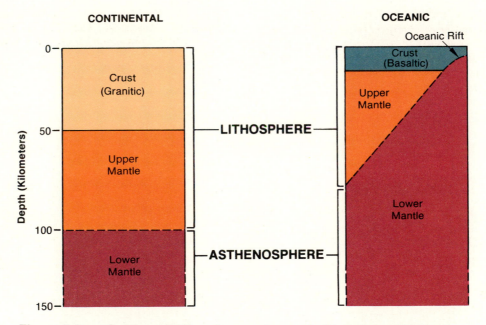

Figure 5.2 Continental lithosphere and oceanic lithosphere differ in thickness and composition. The lithosphere includes the crust and part of the upper mantle. It is divided into rigid plates that move over the underlying asthenosphere, a layer of the mantle that is hotter and weaker than the lithosphere. Continental lithosphere is between 100 and 150 kilometers thick; in contrast, oceanic lithosphere is never more than 100 kilometers thick, and at oceanic rifts, where it is created by the upwelling of the asthenosphere, its thickness is less than 10 kilometers. Continental crust is mostly granitic; oceanic crust consists of basalts and gabbros, both of which form from basaltic magma. (From "The Rifting Continents," by Enrico Bonatti. Copyright © 1987 by Scientific American, Inc. All rights reserved.)

The geologic phenomena associated with plate boundaries depends on the type of movement that occurs there. These are shown generally in figure 5.1. For example, where two plate margins are moving in parallel but opposite directions a major strike-slip fault occurs, such as the San Andreas fault in southern California. In another case, the Nazca Plate in the western Pacific Ocean is impinging against the western margin of the South America Plate. A deep submarine trough, the Peru-Chile Trench, is formed there because the Nazca Plate is dragged beneath the continental lithosphere of the South America Plate. This process is called *subduction,* and the edges of the plates involved constitute a *subduction zone.*

A schematic diagram of subduction is shown in figure 5.3. As the basaltic oceanic crust and some sea floor sediments are dragged beneath the continental crust into a zone of higher temperature, they melt to form magma, which rises slowly toward the earth's surface. When these molten masses reach the surface, they burst forth in volcanic eruptions such as those that are characteristic of the Andean Mountains from southern Chile to Columbia.

In contrast to a *subduction* or *convergent* plate boundary is the *spreading* or *divergent* boundary. The best examples of these—such as the Mid-Atlantic Ridge in the Atlantic Ocean and the East Pacific Rise in the Pacific Ocean (fig. 5.1)—are found in the ocean basins. In both cases, two oceanic plates are moving away from each other. This spreading process is shown in figure 5.3. The axis of the spreading ridge is ruptured to the base of the crust and thin upper mantle, thereby allowing material from the asthenosphere to ooze to the ocean floor as submarine lava flows. The spreading ridge is off-set by *transform faults* that grade along their strikes into *fracture zones.*

The boundaries of active plates are characterized by many earthquakes. The earthquakes generated along spreading ridges and associated *transform faults* have shallow foci (0–70 km deep). In subduction zones, shallow, intermediate (71–300 km deep), and deep (301–700 km deep) earthquakes are generated, as shown in figure 5.3.

With this background information as a guide, you can now proceed to Exercise 24.

References

Bonatti, Enrico. 1987. The rifting continents, *Scientific American,* 256: 96–103.

Dewey, J. F. 1972. Plate tectonics, *Scientific American,* 226: 56–68.

Lowman, P., Wilkes, K., and Ridky, R. W. 1978. Earthquakes and plate boundaries, *Journal of Geological Education,* 26: 69–72.

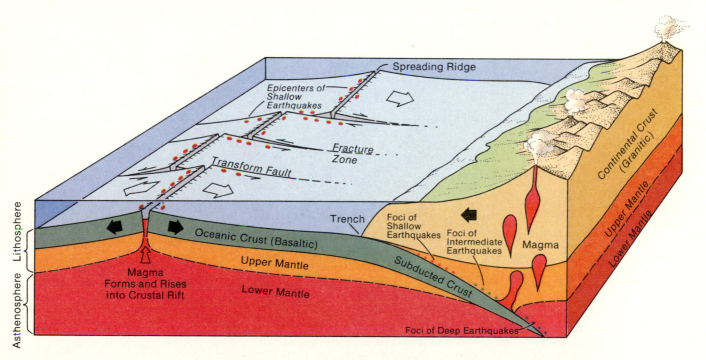

Figure 5.3 Block diagram showing two types of plate margins and the geologic features and events associated with them. A schematic spreading ridge is on the left side and a subduction zone on the right.

Exercise 24. Geologic Phenomena Related to Plate Tectonics

24A. THE NAZCA PLATE, the East Pacific Rise and the Peru-Chile Trench

Locate the Nazca Plate in the eastern Pacific Ocean on figure 5.1. This exercise deals with the western margin of this plate on the East Pacific Rise, and the eastern margin along the west coast of South America.

1. On figure 5.4 sketch the approximate boundary of the Nazca Plate in the Pacific Ocean with black pencil, using the distribution of shallow earthquake epicenters shown by red dots. Use figure 5.1 as a guide, and remember that only a rough boundary can be derived from the earthquake data alone.

2. Based on earthquake data alone, why is the western margin of the Nazca Plate along the East Pacific Rise more likely to be a divergent plate boundary than a convergent one?

3. What pronounced submarine feature along the eastern margin of the Nazca Plate shown in figure 5.1 indicates that it is a convergent plate boundary?

4. Why does the pattern of the distribution of shallow, intermediate, and deep earthquakes in the coastal region of South America add additional support to the idea that the western margin of the Nazca Plate lies in a subduction zone along a convergent plate boundary?

5. Figure 5.5 is a more detailed map of part of the Nazca Plate. The configuration of the East Pacific Rift is shown with detail that you were unable to incorporate in the rough boundary you drew in answer to question 1. The East Pacific Rift is shown in figure 5.5 as bands of color between transform faults and associated fracture zones. The width of each band is drawn to the scale of the map, and represents the amount of spreading (separation) during the last 1 million years.

   a) Using the map scale of figure 5.5, determine the amount of spreading on the East Pacific Rift along the south side of the Wilkes Fracture Zone during the last 1 million years.

   b) Convert your answer to question 5a to centimeters per year (cm/yr). (Recall that 1 kilometer = 1,000 meters, and 1 meter = 100 centimeters.)

Figure 5.4 Epicenters of shallow (0-70 km), intermediate (71-300 km), and deep (301-700 km) earthquakes in part of the eastern Pacific Ocean and parts of North and South America. Shallow earthquake epicenters are shown by red dots, intermediate epicenters by green dots, and deep earthquakes by blue dots. All earthquakes shown on this figure occurred between July 1963 and December 31, 1972, except quakes of 8.0 or more on the Richter scale that occurred between 1897 and 1972 and are shown by a circle of the appropriate color. All ocean depth contours are in fathoms. (From Tarr, A. C., 1974. *World Seismicity Map,* U.S.G.S.)

MIDDLE AMERICA TRENCH

PUERTO RICO

Equator

PACIFIC

RISE

EAST

PERU-CHILE TRENCH

30°

30°

Scale at Equator

1000   500   0   1000   2000

Kilometers

120°

90°

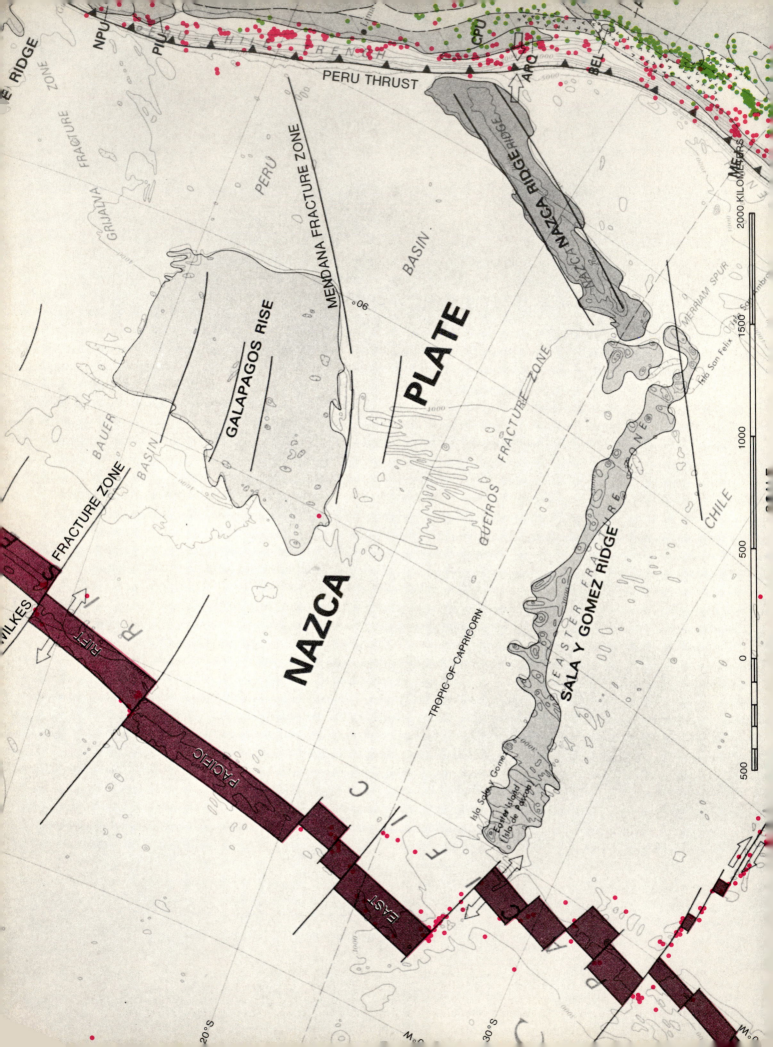

24B. THE HAWAIIAN ISLANDS, the Hawaiian Ridge

The Hawaiian Islands consist of a northwesterly trending chain of islands. Only the largest of these, Hawaii, is an active volcano whose eruptive history you are already familiar with from a previous exercise in this manual. The other islands in the chain are also volcanic in origin, but they ceased erupting some millions of years ago, and have been severely eroded by wave action and intense surface runoff. The Hawaiian Islands are, in fact, part of a longer chain of extinct volcanic islands that comprise the Hawaiian Ridge (fig. 5.6).

It has been postulated that the alignment of the extinct volcanoes forming the Hawaiian Ridge was formed as follows. A *"hot spot,"* whose latitude and longitude has remained fixed over many millions of years, lies in the upper asthenosphere beneath the island of Hawaii. This hot spot is the source of heat that produces the magma that feeds the volcanoes on Hawaii. The oceanic lithosphere has been moving northwesterly over this hot spot in a more or less straight line, and as each nonvolcanic part of the oceanic floor is carried over the hot spot by the conveyor action of the asthenosphere, a new volcano is born. Evidence in support of this hypothesis lies in the absolute dates of old lavas along the Hawaiian Ridge. These lavas are successively older the farther they are from the island of Hawaii.

1. The map of figure 5.6 shows the Hawaiian Ridge, and the absolute dates of lava along it, printed in bold black numbers that represent millions of years before the present. Hawaii is, of course, an active volcano, so the lava from it is zero years old. The lava on Niho Island is 7 million years old. Thus, according to the hot spot hypothesis, Niho Island was once an active volcano standing where the island of Hawaii stands today.

   a) Determine by simple arithmetic the rate of movement of the oceanic lithosphere as it moved over the Hawaiian hot spot. Figure the rate of movement in cm/yr using the distances from Hawaii to each of the three dated lavas on the map of figure 5.6.

   b) Do the rates of movement based on the three dates indicate that the movement has been constant, or variable?

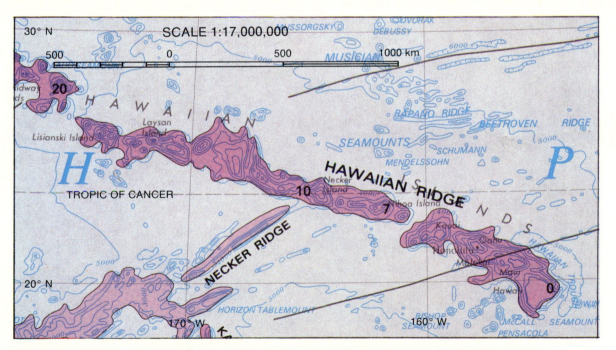

Figure 5.6 Map of the Hawaiian Ridge in the Pacific Ocean. (From the *Preliminary Tectonostratigraphic Map of the Circum-Pacific Region,* 1985, copyrighted and published by the American Association of Petroleum Geologists, Post Office Box 979, Tulsa, Oklahoma 74101.)

Figure 5.5 Map of part of the Nazca Plate. Epicenters of shallow earthquakes are shown in red; intermediate earthquakes in green; and deep earthquakes in blue. (Compiled from the *Plate Tectonic Map of the Circum-Pacific Region,* 1982; and the *Preliminary Tectonostratigraphic Terrane Map of the Circum-Pacific Region,* 1985. Both maps are published by the American Association of Petroleum Geologists, Post Office Box 979, Tulsa, Oklahoma 74101.)

24C. ATOLLS, French Polynesia

An *atoll* is an oceanic island that, in map view, appears as a narrow strip of land with low relief that forms a closed loop. Inside the loop is a shallow lagoon. The loop itself may contain gaps that allow access by ships from the surrounding deep ocean to the lagoon. An atoll is made chiefly of *coral reef*, but it also contains assorted other marine organisms, such as algae and sea shells. Some of the coral may be weathered and eroded by wave action to form coral sand.

Coral is a marine animal that thrives in warm shallow waters of the world's oceans, most notably in the equatorial regions of the Pacific Ocean. Reef-forming corals grow in colonies that attach themselves to the shallow sea floor in tropical and subtropical climatic zones. They live in water that is no more than 50 meters deep, is relatively free of sediments, penetrated by sunlight, and has an abundant food supply of small marine organisms.

In 1842, Charles Darwin proposed a theory for the origin of atolls that he based on his observations of the tropical islands during the voyage of the *Beagle* through the equatorial waters of French Polynesia. Darwin recognized three stages in the evolution of atolls in the islands around Tahiti. These stages are shown in figure 5.7.

Stage I consists of a newly formed volcanic island surrounded by a *fringing coral reef.*

Stage II is reached after the volcanic peak has been eroded by runoff and wave action, and has begun to subside beneath the sea. As the island subsides, the corals of the fringing reef die because the water becomes too deep for their survival. However, the remaining mass of dead coral forms a platform on which new corals establish themselves continuously. This process allows the upward growth of the reef to maintain pace with the sinking of the island. The fringing reef now becomes a *barrier reef,* and a shallow lagoon develops between it and the shore of the volcanic island. The low-lying surface of the barrier reef also may be colonized by vegetation.

[*Continued on p. 188*]

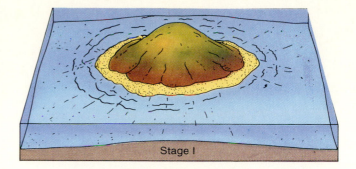

Stage I

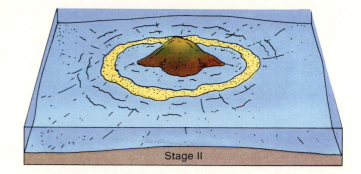

Stage II

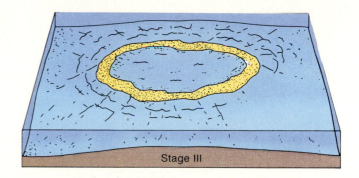

Stage III

Figure 5.7   The three stages in the evolution of atolls as proposed by Charles Darwin. See Exercise 24C for a description of each stage. (From Carla W. Montgomery, *Physical Geology.* Copyright © 1987 Wm. C. Brown Publishers, Dubuque, Iowa. All Rights Reserved. Reprinted by permission.)

24C. *Continued*

Stage III is reached when the eroded remnants of the volcanic peak disappear beneath the sea through continued surface erosion and subsidence of the sea floor. The corals continue their upward growth, and ultimately encircle the enclosed lagoon.

Darwin's theory was challenged by others on the grounds that no mechanism was known that could cause a volcanic island in the middle of the ocean to sink. These critics proposed instead that the evolution of an atoll was due to a rising sea level, not a sinking island. With the advent of plate tectonics, however, Darwin has been vindicated. Volcanic islands are formed by submarine eruptions along spreading plate boundaries. As spreading continues, a volcano is carried on the oceanic crust to deeper water.

The rising sea level proponents of atoll evolution were not totally wrong, however. Sea level has fallen and risen during past geologic time in response to the waxing and waning of Pleistocene ice sheets, and these sea level fluctuations account for some *dead* coral reefs that now lie *above* sea level on the flanks of some volcanic islands.

The modern theory of atoll formation, therefore, calls for a subsiding volcanic island on which sea level fluctuations over geologic time have been superimposed.

1. Figures 5.8, 5.9, and 5.10 represent the three stages in atoll evolution as proposed by Darwin. What are the main identifying characteristics of each that can be observed in the photographs?

2. Why are there gaps in the fringing reef of figure 5.8?

3. What kind of rock would you expect to be encountered by a drill that penetrated the coral and sediments in the center of the lagoon in figure 5.10?

4. Assume that another hole were drilled somewhere on the atoll itself in figure 5.10.
   a) Would the thickness of the coral penetrated by the drill be greater or less than the thickness encountered by the drill in the center of the lagoon?
   b) Sketch a cross-section through the center of the atoll in figure 5.10. Exaggerate the vertical scale and show the following:
      (1) The foundered volcanic island remnant.
      (2) Sea level.
      (3) Coral reef material above and below sea level.
      (4) The location of the two drill holes: A, through the lagoon; B, through the atoll itself.

Figure 5.8   Moorea Island in the Society Islands of French Polynesia in the equatorial Pacific Ocean, an example of Stage I in the evolution of an atoll. (With the permission of and copyrighted by Erwin Christian.)

Figure 5.9   Bora-Bora Island in the Society Islands of French Polynesia in the equatorial Pacific Ocean, an example of Stage II in the evolution of an atoll. (With permission of and copyrighted by Erwin Christian.)

Figure 5.10   Aratica Island in the central Tuamotas of French Polynesia, an example of Stage III in the evolution of an atoll. (With the permission of and copyrighted by Erwin Christian.)

Notes

I Maps
 - map scales & conversions
 - locations using long/lat. & township/range

II. Topographic maps
 - contours, contour interval, relief
 - rules for constructing contour maps (p. 51)
 - construct simple topo maps from a map of given elevations
 - Draw a topographic profile from a topo map.

III fluvial Geomorphology
 - Formation of alluvial fans
 - Lateral migration of streams
 - Formation & evolution of oxbow lakes
 - Gradient, capacity & competence

IV Groundwater
 - Recharge, discharge
 - Effluent & Influent Streams
 - Identify basic features of karst topography
 - Relationship between watertable & sinkholes
 - Use flowlines to predict direction of groundwater flow

V) Glaciation
 - New snow, old snow, firn, glacial ice
 - Snow line, net mass balance of glacier
 - Cirque, arete, tarn, hanging valley, drumlin, moraine

VI Structural Geology
 - Strike + dip
 - angle + direction of attitude from geologic map.
 - Anticline and syncline
 - Faults + how to name them (normal reverse thrust).
 - Offsetting of beds due to faulting
 - ages of rocks from a geologic map
 - Map symbols

5	10
10	10
8	10
10	10
117	150
15	15
29	30
30	30
30	30
10	10
10	10
15	15
289	330

$$100 \cdot \frac{289 + x}{2} = .90 \%$$

$$(.9 \times 2) \overline{)\, 289}$$

$$\frac{480}{} = .9$$

$$\frac{x + 289}{480} = .9$$

Notes

Notes

Notes